Motive powe

EM

Colin J. Marsden

LONDON

IAN ALLAN LTD

First published 1982

ISBN 0 7110 1165 6

Published by Ian Allan Ltd, Shepperton, Surrey; and printed by Ian Allan Printing Ltd at their works at Coombelands in Runnymede, England

Introduction

In this the second volume in the Motive Power Recognition series, BR electric multiple units are dealt with. At the end of 1980 there was a total of 2,204 units, formed of 7,718 coaches in capital stock, conforming to the basic EMU requirements.

Many of the EMU types and classes are of the same basic design, and this volume has been produced to try and assist railway enthusiasts in identifying various classes, sub-classes or types. Also included is a short technical section on each class, which will be of interest to the technical enthusiast rather than the train spotter.

Electric multiple unit formed trains are most widely used on the Southern Region where some 90% of all passenger services are in their hands. The lines out of Liverpool Street on the Eastern Region are the second largest users, followed by the London Midland Region, Scottish Region (Glasgow suburban network), and the GN electrified lines out of London's Kings Cross.

Numerical class numbering for EMU stock was introduced during the 1970s, with class numbers denoting operating differences. Units in the 3XX range being fitted for AC overhead power collection, 4XX range numbers were allocated to Southern Region DC units, and 5XX numbers were allocated to all other DC units.

During the late 1970s a start was made to renumber EMU stock with the class prefix computer system, but up to early 1981 less than 20% of units have been renumbered. All stock commissioned after 1975 has carried its new number from the offset. Refurbished SR Class 411 units have been given their new numbers while in Swindon Works.

The present livery for EMU stock is standard rail blue or blue and grey, with full yellow ends, many of the older units in all blue livery, are undergoing refurbishing programmes, and will emerge in blue and grey livery. Unit numbers are applied to each end of fixed sets, usually above window height, or under the driver's side nose window. Some DC units do not carry fixed set numbers, and are recognised by coach numbers.

This volume contains technical data and photographic illustrations of all classes of Electric Multiple Unit in traffic at the end of 1980, together with many of the departmental and service units that conform to EMU requirements. It is hoped that readers of this volume will find it useful when trying to identify or recognise various Electric Multiple Unit types.

I would like to thank many of my friends and members of BR who have given assistance with material for this volume, and Mr K. Loughland for assistance with some of the technical data.

The editor of this volume would like to be notified of any major detail differences that have been omitted from this book by sending them, together with a photograph if possible to: The Editor, Motive Power Recognition, Ian Allan Ltd, Terminal House, Shepperton TW17 8AS

Colin Marsden
Surbiton
March 1981

Left: *Class 405 unit No 4751 leaves Clapham Junction in August 1981.* Colin J. Marsden

Front cover, top: *Class 312/0 No 312.015 (now 312.715) arrives at Kings Cross.* Colin J. Marsden

Front cover, bottom left: *Front end of refurbished Class 411 unit.* Colin J. Marsden

Front cover, bottom right: *Front end of Class 508 unit.* Colin J. Marsden

Back cover, top left: *Front end of LM Class 304/3 unit.* Brian Morrison

Back cover, top right: *Front end of ER Class 302 unit.* Brian Morrison

Back cover, bottom: *Waterloo & City Railway motor car at Alton.* Colin J. Marsden

Electric Multiple Unit Types and Classification

Class	*Type Code*	*Formation*	*Region of Allocation*
302		BDTS-MBS-TC-DTS	Eastern
303		BDTS-MBS-DTS	Scottish
304		BDTS-MBS-TC-DTBS	Midland
305/1		BDTS-MBS-DTS	Eastern
305/2		BDTS-MBS-TC-DTS	Eastern
306		DMS-TBS-DTS	Eastern
307		DTBS-MS-TC-DTS	Eastern
308/1		BDTS-MBS-TC-DTS	Eastern
308/2		BDTS-MLV-TC-DTS	Eastern
308/3		BDTS-MBS-DTS	Eastern
308/4		BDTS-MBS-TC-DTS	Eastern
309/1 & 4		DMBS-TS-TC-BDTS	Eastern
309/2		BDTC-MBS-TG-DTC	Eastern
309/3		BDTC-MBS-TS-DTC	Eastern
310		BDTS-MBS-TS-BDTC	Midland
311		DTS-MBS-BDTS	Scottish
312		BDTS-MBS-TS-DTC	Eastern/Midland
313		DMS-TC-BDMS	Eastern
314		DMS-TS-DMS	Scottish
315		DMS-TS-TS-BDMS	Eastern
370	APT	DTS-TS-TRSB-TU-TF-TBF-M	Scottish
405	4SUB	MBSO-TS-TSO-MBSO	Southern
410	4BEP	DMBS-TC-TRB-DMBS	Southern
411	4CEP	DMBS-TC-TS-DMBS	Southern
411*	4CEP	DMS-TBC-TS-DMS	Southern
414	2HAP	DMBS-DTC	Southern
415/1	4EPB	MBSO-TS-TSO-MBSO	Southern
415/2	4EPB	MBSO-TS-TS-MBSO	Southern
416	2EPB	DMBS-DTS	Southern
418	2SAP	DMBS-DTS	Southern
419	1MLV	MLV	Southern
420	4BIG	DTC-MBS-TRBS-DTC	Southern
421	4CIG	DTC-MBS-TS-DTC	Southern
423	4VEP	DTC-MBS-TS-DTC	Southern
427	4VEG	DTC-MBS-TS-DTC	Southern
430	4REP	MSO-TRB-TBF-MSO	Southern
485	4VEC	MBSO-TSO-DTSO-MBSO	Southern
486	3TIS	MBSO-TSO-MBSO	Southern
487		DMBS and TS	Southern
491	4TC	DTSO-TF-TBS-DTSO	Southern
501		DMBS-TS-DTBS	Midland
503		DMBS-TS-DTS	Midland
504		DMBS-DTS	Midland
506		DMBS-TS-DTS	Midland
507		BDMS-TS-DMS	Midland
508	4PER	DMS-TS-TS-BDMS	Southern

Electric Multiple Unit Coach Classification

BDMS	Battery driving motor second
BDTC	Battery driving trailer composite
BDTS	Battery driving trailer second
DMBS	Driving motor brake second
DMS	Driving motor second
DTBS	Driving trailer brake second
DTC	Driving trailer composite
DTS	Driving trailer second
DTSO	Driving trailer second open
M	Motor
MBS	Motor brake second
MBSO	Motor brake second open
MLV	Motor luggage van
MS	Motor second
MSO	Motor second open
TBC	Trailer brake composite
TBF	Trailer brake first
TBS	Trailer brake second
TC	Trailer composite
TF	Trailer first
TG	Trailer griddle
TRB	Trailer buffet unclassified
TRBS	Trailer restaurant buffet second
TRSB	Trailer restaurant second buffet
TS	Trailer second
TSO	Trailer second open
TU	Trailer unclassified

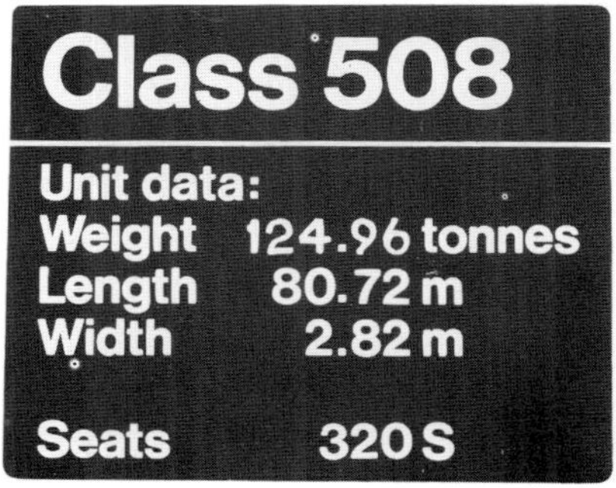

Right: *Unit Data Panel. Panels similar to this will be found on all units and coaches in traffic, all giving the same basic information. Coach Data Panels are of similar appearance but give coach code rather than class as heading. Data is self explanatory.* Colin J. Marsden

Southern Region Electric Multiple Unit Types

Southern Region electric multiple units are usually referred to by the SR letter code system (listed below) rather than the BR class numbers. SR units are also referred to by years of manufacture, ie 36, 51, 57, 63, 66, and 72. These dates do not refer to actual years of construction, but the years of equipment manufacture or design.

Unit Types

SUB	Suburban unit	1936 stock
BEP	Buffet electro-pneumatic	*1957 stock
CEP	Corridor electro-pneumatic	*1957 stock
HAP	Half lavatory electro-pneumatic	1957 stock
EPB	Electro-pneumatic brake	†1951 stock/1957 stock
SAP	Second electro-pneumatic	1951 stock
MLV	Motor luggage van	1957 stock
BIG	Buffet intermediate gangway	1963 stock
CIG	Corridor intermediate gangway	1963 stock
VEP	Vestibule electro-pneumatic	1963 stock
VEG	Vestibule electro-Gatwick	1963 stock
REP	Restaurant electro-pneumatic	1966 stock
TC	Trailer control	1966 stock
PER	SR high density stock	1972 stock

* Some units are of 1951 type stock.
† 1951 stock = 50XX, 51XX, 54XX, 56XX, 57XX series units.
1957 stock = 53XX series units.

Multiple Operation of SR Stock

Due to operating difficulties involving brakes, control and lighting systems, SR units are not able to be coupled universally, therefore a simple grouping system has been devised to assist operating staff.

Unit Groups

GROUP A: 4SUB (Class 405), CM&EE stores units No 022-024, Instruction unit No 055
GROUP B: NOT USED
GROUP C: NOT USED
GROUP D: 4EPB (Class 415), 2EPB (Class 416), 2SAP (Class 418), De-icing units No 001-013
GROUP E: 4BEP (Class 410), 4CEP (Class 411), 2HAP (Class 414), 4BIG (Class 420), 4CIG (Class 421), 4VEP (Class 423), 4VEG (Class 427, MLV (Class 419)
CLASS F: 1972 design High Density stock (Class 508)

Units of Group D & E can inter couple, providing that 50% of units are of Group E, otherwise only units of the same group can be fully coupled together.

Multiple operation of 4REP (Class 430) tractor units

4REP (Class 430) units, due to their large power requirements, are not permitted in multiple operation with any other powered unit, except a 2-car or MLV unit, but this system of operation is not often seen. If for any reason a REP unit has to be coupled or work with another powered unit the motors of one unit are 'cut out'.

Right: *Class 423 unit No 7876 passes Wandsworth Common on 13 August 1981.*
Colin J. Marsden

7876
7876
34

Class 302

Class type: AM2
Unit numbers: 201-312
Computer numbers: 302201-302312
Introduced: 1959
Built by: BR Doncaster/York
Equipment manufacturer: English Electric
Area of use: Fenchurch Street lines
Weight of unit: 155ton
Length of unit: 266ft 7in
Height: 13ft 0½in
Width: 9ft 3in
Brake type: Auto air/electro-pneumatic
Nominal supply voltage: 6.25kV and 25kV AC
Type of power collection: Overhead
Traction motor type: English Electric
Available hp: 768
Number of traction motors: 4
Gear ratio: 65:17
Max unit speed: 75mph
Bogie type: Gresley design
Coupling type: Buck-eye
Coupling restriction: Eastern EMU stock
Formation and seating: BDTS 80 second; MBS 96 second; TC 19 first/60 second; DTS 108 second

Below: *To cater for the electrification scheme of the Fenchurch Street to Shoeburyness and Tilbury lines a batch of 112 4-car sets was built from 1959, each unit seating 344 second and 19 first class passengers. In this view No 302259 approaches Leigh-on-Sea with the 15.45 Shoeburyness to Fenchurch Street working on 28 September 1980* Les Bertram

Above: *From the mid-1970s the use of the headcode system in common with most other EMU classes was dispensed with. It is usual that the headcode blinds are wound to the white blank or O position, as depicted in this view of unit No 265 at Shoeburyness on 15 April 1980.* R. E. Ruffell

Below: *The power car is in the middle of the formation and can be recognised at all times as the pantograph is fixed on the roof of this car, above the guard's brake van. Traction motors are mounted on the bogies under this coach and associated electrical equipment is slung between. This illustration shows MBS No E61196 of unit No 280 at Ilford.* B. Denton

Above: *Front end layout of Class 302 .(**1**) Multiple unit control jumper, (**2**) Main reservoir pipe, (**3**) Brake pipe, (**4**) Multiple unit control jumper, (**5**) Red tail marker lights, (**6**) Warning horns, (**7**) Buck-eye coupling, (**8**) AWS receiver equipment, (**9**) Release chain for buck-eye coupling. Unit No 206 in this photograph stands at Southend Victoria during 1976.* Brian Morrison

Above: *TC (Trailer Composite) coaches of Class 302 units are laid out for first and second class seating. Nearest the camera the six second class seating bays will be seen (seating 60 passengers). At the far end three first class seating bays can be seen (seating 19 first class passengers). In between the first and second class seating is situated two lavatories. TC No E70072 of unit No 212 stands at Romford during the early part of 1981.* K. Grafton

Below: *A unit refurbishment programme was authorised during the late 1970s to improve seating and facilities on these 20-year old units. This view of a Class 302 driving trailer (ex unit No 204) shows the revised seating arrangement with first class accommodation now in the driving trailer directly behind the driver. During 1981 several units of this class were refurbished in a similar manner. The unit number applied to this demonstration coach was fictitious and only affixed for evaluation purposes.* K. Grafton

Class 303

Class type: AM3
Unit numbers: 001-091
Computer numbers: 303001-303091
Introduced: From 1959
Built by: Pressed Steel
Equipment manufacturer: English Electric
Area of use: Glasgow suburban
Weight of unit: 127ton
Length of unit: 199ft 6in
Height: 12ft 8in
Width: 9ft 3in
Brake type: Auto air/electro-pneumatic
Nominal supply voltage: 25kV AC
Type of power collection: Overhead
Traction motor type: AEI (MV)
Available hp: 828
Number of traction motors: 4
Gear ratio: 70:17
Max unit speed: 75mph
Bogie type: Gresley design
Coupling type: Buck-eye
Coupling restriction: ScR EMU types
Formation and seating: DTS 83 second; MBS 70 second; BDTS 83 second

Left: *Unit No 012, one of 91 3-car units built between 1959 and 1961 by Pressed Steel Ltd for use on the North Clyde electrification, departs from Glasgow Central on 15 November 1976. The livery carried by this example is BR standard rail blue. Note this unit also has the old wrap round driver's windows, which have now been replaced by the flat fronted type.* Brian Morrison

Above: *Front end arrangement of Class 303 unit: (**1**) Multiple unit control jumper cable, (**2**) Main reservoir pipe, (**3**) Brake pipe, (**4**) Multiple unit control jumper socket, (**5**) Red tail indicator lights (two), (**6**) Warning horns, (**7**) Buck-eye coupling, (**8**) AWS receiver equipment. Unit No 303016 stands at Glasgow Shields Road Depot during 1978. This front end arrangement also applies to Class 311 units.* T. Noble

Above: *When introduced units were painted in electric blue, but from the early 1970s standard BR blue has been applied, until the late 1970s when some units have emerged in blue and grey livery. Unit No 002 departs from Glasgow in this illustration.* B. Denton.

Below: *Following the TOPS (Total Operation Processing System) numbering sequence all units have had the class prefix added to the unit number. In this view taken at Gourock during 1980 unit No 081 is still awaiting the prefix to be added.* Brian Morrison

Above: *Following the South Clyde electrification of 1967 a further batch of identical units was constructed, this time by the firm of Cravens Ltd, these units were classified as 311 and externally the numbers are the only recognition factor between the two different types. Here No 303094 stands next to No 311102 at Glasgow Central on 14 September 1980.* C. P. Boocock

Below: *Units are fitted with air operated sliding doors controlled by a master switch in the guard's brake van. This view of the DTS of No 303027 at Hyndland Depot shows one of the double leaf doors in the open position. The 'door open' button can be seen towards the upper part of the body next to the door. The DTS cars of these units carry little underframe equipment with only brake equipment being fitted (other side of coach).* T. Noble

Class 304

Class type: AM4
Unit numbers: 001-045
Computer numbers: 304001-304045
Introduced: 1960
Built by: BR Wolverton
Equipment manufacturer: AEI Traction Ltd
Area of use: Euston-Carlisle and branches
Weight of unit: 152-153ton
Length of unit: 246ft 6½in
Height: 12ft 4½in
Width: 9ft 3in
Brake type: Auto air/electro-pneumatic
Nominal supply voltage: 25kV AC
Type of power collection: Overhead
Traction motor type: BTH
Available hp: 828
Number of traction motors: 4
Gear ratio: 17:70
Max unit speed: 75mph
Bogie type: Gresley design
Coupling type: Buck-eye
Coupling restriction: Within LMR type
Formation and seating: BDTS 80 second; MBS 96 second*; TC 19 first/60 second; DTBS 82 second

*MBS coaches in unit Nos 016 to 045 seat 72 second.

Below: *Following the electrification of the LMR line between Manchester and Crewe in 1960, a batch of 15 4-car units was constructed by BR at Wolverton and later classified as 304. The units are of the classic 1960 construction style and are almost identical to many other AC and DC units. Following the success of these units a further batch of 20 was constructed in early 1961 and one of these units No 022 is seen in this illustration arriving at Bescot on 10 July 1980.* L. Bertram

Above: *The main difference between the batches is in the seating configuration, and no external difference exists to help with recognition. A further batch to 10 units was constructed during 1961 which completed the class. This view of No 037 with BDTS No M75869 leading was taken at Great Bridgford junction on 21 July 1978.*
J. Chalcraft

Below: *Two units, one with revised class prefix number stand at Longsight Depot on 8 June 1980. Usually Class 304 units work north of Rugby but very seldom travel north of Weaver junction on the main line, their work usually being confined to the Birmingham, Crewe and Manchester areas.* R. G. Fox

Above: *Class 304 Front end layout: (**1**) Multiple unit control jumper, (**2**) Main reservoir pipe, (**3**) Brake pipe, (**4**) Multiple unit control jumper socket, (**5**) Rubbing plate, (**6**) Buck-eye coupling, (**7**) AWS receiving equipment, (**8**) Red tail indicator lights. No 304040 was photographed at Rugby during 1979.* Brian Morrison

Above: *All power equipment is collected and transformed into traction current in the Motor Brake Second Coach, a standard single arm pantograph being fitted above the guard's brake van. The boxes between the bogies contain control, power and auxiliary equipment.* Colin J. Marsden

Below: *In this view TC (Trailer Composite) No M70490 of unit No 023 clearly shows the Gresley design bogies and empty underframe. On these cars seating is arranged in the 3+2 Saloon style for second class passengers, and in three compartments for first class passengers, accommodation being separated by two lavatories.*
Colin J. Marsden

Class 305

Class number: 305/1, 305/2
Class type: AM5
Unit numbers: 401-455 (305/1)
501-519 (305/2)
Computer numbers: 305401-305455 (305/1)
305501-305519 (305/2)
Introduced: 1960
Built by: BR York (305/1)
BR Doncaster (305/2)
Equipment manufacturer: English Electric (AEI Ltd)
Area of use: Liverpool St-Enfield, Chingford (305/1)
GE Outer Suburban (305/2)
Weight of unit: 119ton (305/1)
153ton (305/2)
Length of unit: 200ft 1in (305/1)
266ft 8½in (305/2)
Height: 12ft 6½in
Width: 9ft 3in
Brake type: Auto-air/electro-pneumatic
Nominal supply voltage: 6.25kV and 25kV AC
Type of power collection: Overhead
Traction motor type: English Electric
Available hp: 800
Number of traction motors: 4
Gear ratio: 65:17
Max unit speed: 75mph
Bogie type: Gresley design
Coupling type: Buck-eye
Coupling restriction: Eastern EMU stock
Formation and seating: BDTS 94 second; MBS 84 second; DTS 94 second (305/1)
BDTS 80 second; MBS 96 second; TC 19 first/60 second; DTS 108 second (305/2)

Below: *Another fleet of the standard 1960 design EMU stock was this fleet of 52 3-car units for use on the Liverpool Street suburban system. These 3-car units seat 272 second class passengers and are classified 305/1 and are currently to be seen over most parts of the Liverpool Street suburban system. No 438 departs from Tottenham Hale during 1976.* Brian Morrison

Above: *Units are mounted on a Gresley designed bogie of non-standard type, and are formed with the MBS (Motor Brake Second) in the middle of the formation, all underframe mounted power and auxiliary equipment is slung between the two power bogies. The MBS of unit No 415 was photographed at Bethnal Green during 1980.*
Colin J. Marsden

Below: *DTS (Driving Trailer Second) of unit No 415. Bodyside windows are of the non-opening type, but doors are fitted with standard droplights. Seating is of the 3+2 formation with 18 doors per car, the only underslung equipment is for brake control.*
Colin J. Marsden

Above: *From the late 1970s some units were emerging from works painted in standard inter-city livery of blue and grey, which is depicted in this view of No 438 entering Bethnal Green station. The unit is passing over one of the AWS (Automatic Warning System) track mounted magnets, situated 200yd on the approach side of all signals. The train receiver is mounted between the two front wheels of the unit.* Colin J. Marsden

Below: *In this view of No 431 leaving Ponders End with the 17.40 Hertford East to Liverpool Street working during June 1980 the BDTS end is leading. The battery boxes and air reservoirs can be seen mounted on the underframe.* L. Bertram

Above: *Nineteen units are classified 305/2 and carry running numbers 501-519. The outward appearance from the front end is identical to the 305/1 fleet, but these units are formed of four coaches, the additional coach being a TC (Trailer Composite) giving some first class accommodation to the train. Unit No 509 in this illustration departs from Bishops Stortford on 3 August 1980.* Andrew French

Below: *These 19 305/2 units also differ from their 3-car sisters in that the MBS (Motor Brake Second) and DTS (Driving Trailer Second) coaches are of an all compartment type. Unit No 504 here stands at Bishops Stortford during August 1980. These units are scheduled for refurbishing during the early 1980s.* M. J. Collins

Class 307

Class type: AM7
Unit numbers: 101-132
Computer numbers: 307101-307132
Introduced: 1956 rebuilt 1960-62
Built by: BR Eastleigh
Equipment manufacturer: English Electric
Area of use: GE outer suburban
Weight of unit: 155ton
Length of unit: 265ft 8½in
Height: 13ft 0½in
Width: 9ft 3in
Brake type: Auto air/electro-pneumatic
Nominal supply voltage: 6.25kV and 25kV AC
Type of power collection: Overhead
Traction motor type: English Electric
Available hp: 700
Number of traction motors: 4
Gear ratio: 74:16
Max unit speed: 75mph
Bogie type: ET, ED, and Gresley
Coupling type: Buck-eye
Coupling restriction: Eastern EMU stock
Formation and seating: DTBS 84 second; MS 120 second; TC 19 first/60 second; DTS 80 second

Below: *These 32 4-car Eastleigh built units were originally equipped for 1,500v DC operation, but were subsequently converted for AC working from 1960. Units can usually be found working the Liverpool Street-Southend (Victoria) line. Unit No 118 in this plate approaches Harold Wood on 6 May 1978.* Brian Morrison

Below: *Class 307 front end layout:* ***(1)*** *Multiple unit control jumper,* ***(2)*** *Main reservoir pipe,* ***(3)*** *Brake pipe,* ***(4)*** *Multiple unit control jumper socket,* ***(5)*** *Red tail indicator light,* ***(6)*** *Rubbing plate,* ***(7)*** *Buck-eye coupling,* ***(8)*** *Air horns. Unit No 307115 was photographed at Liverpool Street during 1980.* Colin J. Marsden

Above: *MS (Motor Second) cars are marshalled between the DTBS (Driving Trailer Brake Second) and the TC (Trailer Composite) coaches. Power is collected from the overhead live wire via a pantograph mounted on the roof of the DTBS, and passes through a transrector system prior to being transferred by means of a power train line to the MS vehicle, In this illustration unit No 126 passes Brentwood Bank.*
Brian Morrison

Below: *The TC (Trailer Composite) vehicles of Class 307 units are almost identical in appearance to that of Class 302s. The main recognition factor being stepboards, that are of the solid strip type on Class 307s, whereas on Class 302s individual boards by each door are provided. TC No E70025 of unit No 125, painted in inter-city livery stands at Romford during early 1981.* K. Grafton

Above: *Units of Class 307 are very similar in appearance to Class 302 stock but are recognisable by the following factors. Class 307 units have their pantograph mounted upon the roof of the DTBS (Driving Trailer Brake Second) where on Class 302 units the pantograph is mounted upon the MBS (Motor Brake Second) which is an intermediate car. 307 units only have one red tail indicator lamp in place of two on 302 stock, cab front windows are also slightly narrower on Class 307 stock. If the two classes are compared these minor details will become apparent. Unit No 101 in this illustration passes Bethnal Green during the winter of 1979.* Brian Morrison

Below: *This plate has been included to show the front end difference between the old Class 306 stock (now withdrawn) and the more modern Class 307 type. Buck-eye couplings and rubbing plates were fitted to 307 stock while on the 306 stock only built about six years earlier, standard screw shackle couplings were fitted. All nose end connections on Class 307 stock are fitted at waist height whilst on 306 stock control jumpers were at waist height and air connections at ground level.* L. Bertram

Class 308

Class numbers: 308/1, 308/2, 308/3, 308/4
Unit numbers: 133-165 (308/1)
317-321 (308/2)
453-455 (308/3)
313-316 (308/4)
Computer numbers:
308113-308165 (308/1)
308317-308321 (308/2)
308453-308455 (308/3)
308313-308316 (308/4)
Introduced: 1961 (308/4 converted 1971 from 308/2)
Built by: BR at York
Equipment manufacturer: English Electric
Area of use: GE and Fenchurch Street lines
Weight of unit: 154ton (308/1)
153ton (308/2)
119ton (308/3)
154ton (308/4)
Length of unit: 268ft $8\frac{1}{2}$in (308/1, 308/2, 308/4)
189ft $9\frac{1}{2}$in (308/3)
Height: 13ft $0\frac{1}{2}$in (max)
Width: 9ft 3in
Brake type: Auto air/electro-pneumatic
Nominal supply voltage: 6.25kV and 25kV AC
Type of power collection: Overhead
Traction motor type: English Electric
Available hp: 800
Number of traction motors: 4
Gear ratio: 65:17
Max unit speed: 75mph
Bogie type: Gresley design
Coupling type: Buck-eye
Coupling restriction: Eastern EMU stock
Formation and seating: BDTS 80 second; MBS 96 second; TC 19 first/60 second; DTS 108 second (308/1)
BDTS 80 second; MLV —; TC 19 first/60 second; DTS 108 second (308/2)
BDTS 94 second; MBS 84 second; DTS 94 second (308/3)
BDTS 80 second; MBS 76 second; TC 19 first/60 second; DTS 108 second (308/4)

Below: *The Class 308 series stock was first introduced in 1961 for use on the Colchester, Clacton and Walton lines. Now in the 1980s there are four sub-classes to the 308 series. 308/1 stock being of four coach formation and numbered 133 to 165 soon to be renumbered 308133-308165. All are allocated to Clacton or Ilford sheds. In this plate unit No 144 passes Marks Tay station during 1980.* M. Collins

Above: *All Class 308 stock are of the same basic body design and are built to the classic EMU design of the early 1960s. Unit No 165 with DTS (Driving Trailer Second) leading was photographed near Romford during 1980* B. Denton

Below: *The only MLV (Motor Luggage Vans) formed within fixed electric multiple unit sets are marshalled in the formation of five Class 308/2 units, (Nos 317-321). These units were designated for use on the Tilbury line, and were intended for operation when the passenger flow was light, ie during very early morning or very late in the evening. In this view 308/2 unit No 317 stands at Tilbury Riverside with the MLV second from the back. In the foreground a Class 302 unit can be seen.* John Glover

Above: *A batch of three units was constructed in 1961 at York works fitted with English Electric power/control equipment and formed of only three coaches giving a total seating capacity of only 272 second class passengers. Units are numbered 453 to 455 and are allocated to Ilford Depot usually working with Class 305 units. General body appearance is the same as other batches of 308 units. In this view unit No 454 approaches Bethnal Green during late 1979.* Brian Morrison

Below: *From the early 1970s four of the original MLV coaches were converted into MBS (Motor Brake Second) cars. This view shows the external body appearance on one such car No E62434 of unit No 314. Underframe mounted equipment is for power and control systems. MBS coaches seat 76 second class passengers in five bays. Converted units are classified as Class 308/4.* John Glover

Above: *Class 308/4 4-car units were converted from Class 308/2 units in 1971, and now seat 324 second class and 19 first class passengers, the units can normally be seen operating on the Fenchurch Street to Shoeburyness line, and are allocated to East Ham Electric Multiple Unit shed. Unit No 316 passes Woodgrange Park in this illustration during early 1980, when its train was diverted to Liverpool Street station.* Brian Morrison

Below: *Unit No 314 in this illustration soon to be renumbered 308314 stands at Benfleet Station with an afternoon Fenchurch Street to Shoeburyness service during late 1980. The empty underframe of the DTS (Driving Trailer Second) coach can clearly be seen. DTS cars are always marshalled at the opposite end of the unit to the MBS (Motor Brake Second) coach.* Les Bertram

Class 309

Class numbers: 309/1, 309/2, 309/3, 309/4
Class type: AM9
Unit numbers: 601-604 (309/1)
605/608 (309/4)
611-618 (309/2)
621-627 (309/3)
Computer numbers:
309601-309604 (309/1)
309605-309608 (309/4)
309611-309618 (309/2)
309621-309627 (309/3)
Introduced: 1961
Built by: BR York
Equipment manufacturer: English Electric
Area of use: Liverpool Street-Clacton & Walton
Weight of unit: 172ton (309/1)
168ton (309/2)
167ton (309/3)
172ton (309/4)
Length of unit: 265ft $8\frac{1}{2}$in
Height: 13ft $0\frac{1}{2}$in
Width: 9ft 3in
Brake type: Auto air/electro-pneumatic
Nominal supply voltage: 6.25kV and 25kV AC
Type of power collection: Overhead
Traction motor type: English Electric
Available hp: 1,128
Number of traction motors: 4
Gear ratio: 59:21
Max unit speed: 100mph
Bogie type: Commonwealth
Coupling type: Buck-eye
Coupling restriction: Eastern EMU stock
Formation and seating: DMBS 40 second; TS 64 second; TC 24 first/24 second; BDTS 60 second (309/1)
BDTC 18 first/32 second; MRS 48 second; TC 24 unclassified; DTC 18 first/32 second (309/2)
BDTC 18 first/32 second; MBS 48 second; TG 64 second; DTC 18 first/32 second (309/3)
DMBS 48 second; TS 64 second; TC 24 first/24 second; BDTS 60 second (309/4)

Below: *The 309 series units of medium power output were designed and built in 1961 exclusively for use on the Liverpool Street to Clacton and Walton lines. Class 309 units are currently divided into four sub classes. 309/1 stock when built were only two car sets, but following reorganisation of the line two trailers have recently been added to make four car sets. 309/1 units can be identified from other sub classes by having their power collection pantographs mounted on the DMBS (Driving Motor Brake Second) coach rather than on MBS (Motor Brake Second) coaches. One of the two coach sets No 603 departs from Witham in the summer of 1980.* M. Collins

Above: *The batch designated 309/3 are 4-car sets with MBS (Motor Brake Second) in the middle of the formation. When built these were the first EMU stock to be completely carried on commonwealth bogies, and over the years have proved to be very satisfactory and have given a good ride. Front end gangways are also provided on this stock for use when working in multiple, a feature more commonplace on SR stock. Unit No 621 in this illustration leads the 09.50 Liverpool Street to Clacton past Brentwood Bank on 19 May 1977.* Brian Morrison

Below: *In this illustration TS (Trailer Second) No E70254 of Class 309/3 unit 622 stands at Liverpool Street during early 1981. TS vehicles in 309/3 sets are of the low density 2+2 layout each seating bay being provided with a full size table. Underframe mounted equipment between the bogies consists of only brake and EP systems.* K. Grafton

Above: *MBS (Motor Brake Second) coaches of 309/2 and 309/3 units are fitted with roof mounted pantographs, above the guard's van. Underframe mounted equipment, between the bogies consists of power and control equipment. The fourth seating bay from the end nearest the camera is fitted with a stretcher window, which today is not often used. In this illustration MBS No E61933 of unit No 612 stands at Liverpool Street during early 1981.* K. Grafton

Below: *This plate illustrates the front end differences between Class 309 and 307 stock. The slightly off centre headcode box on the 309 unit is clearly visible in this photograph. With the front end gangway it was necessary to fit air pipes on both sides of the unit to ease coupling and uncoupling procedures. On the left the multiple unit control jumper can be seen, while on the right side of the gangway the multiple unit control jumper socket is to be found (to the right of the air pipe).* K. Grafton

Right: *From the mid-1970s the availability and cost of the curved glass front windows became prohibitive and a revised front window layout was designed. The revised arrangement is clearly shown in this illustration fitted to the unit on the left, while the unit on the right still has the old style windows fitted.* M. Pudduck

DANGER

Class 310

Class type: AM10
Unit numbers: 046-095
Computer numbers: 310046-310095
Introduced: 1965
Built by: BR Derby
Equipment manufacturer: English Electric
Area of use: Euston-Carlisle and branches
Weight of unit: 158ton
Length of unit: 265ft 8½in
Height: 13ft 0½in
Width: 9ft 3in
Brake type: Auto air/electro-pneumatic
Nominal supply voltage: 25kV AC
Type of power collection: Overhead
Traction motor type: English Electric
Available hp: 1,080
Number of traction motors: 4
Gear ratio: 63:17
Max unit speed: 75mph
Bogie type: B4
Coupling type: Buck-eye
Coupling restriction: Within LMR type
Formation and seating: BDTS 80 second; MBS 70 second; TS 100 second; BDTC 26 first/43 second

Below: *From 1965 BR workshops at Derby constructed 50 4-car high density units to work on the Euston Suburban services. Now during 1982 only 49 units remain in traffic, this follows collision damage to one unit. Units are semi-permanently coupled into 4-car sets with power collection being affected by a roof mounted pantograph on the MBS (Motor Brake Second) vehicle. Unit No 070 passes near Wembley with a Rugby-Euston local.* Colin J. Marsden

Above: *Class 310 front end layout: (**1**) Multiple unit control jumper, (**2**) Main reservoir pipe, (**3**) Brake pipe, (**4**) Multiple unit control jumper socket, (**5**) Red tail indicator light, (**6**) Rubbing plate, (**7**) Buck-eye coupling (in lower position), (**8**) Buck-eye coupling release chain, (**9**) Buffer saddle (for extending buffer position when buck-eye coupling is in lower position). Unit No 047 in this illustration displays the old livery, now all units are painted with a full yellow end.* British Railways

Above: *Detail of MBS coach No M62071 of unit No 046. The pantograph is again above the guard's brake van. Electrical equipment is contained in a cupboard inside the van as well as in saddle boxes slung between the two motor bogies. No corridor connection exists between the MBS (Motor Brake Second) and the BDTS (Battery Driving Trailer Second) coach.* British Railways

Below: *From the late 1970s some units have emerged from works overhaul at Wolverton Works painted in standard blue and grey livery. Most of the 'new livery' units also carry the class prefix number. In this illustration of No 310056 arriving at Euston, a modified marker light system has been fixed in the former headcode box, as three white discs are fitted in place of the normal two.* Colin J. Marsden

Driving Trailer coaches are of different designs at each end of the unit. In the upper illustration unit No 072 is photographed from the BDTC (Battery Driving Trailer Composite) end with the first class accommodation behind the driver's cab. The lower illustration of unit No 049 was photographed from the BDTS (Battery Driving Trailer Second) end. This view also shows the buck-eye coupling in the lowered position and buffers in extended position. Both B. Nicolle

Class 311

Class type: AM11
Unit numbers: 092-110
Computer numbers: 311092-311110
Introduced: 1967
Built by: Cravens
Equipment manufacturer: AEI Traction Ltd
Area of use: Glasgow suburban
Weight of unit: 129tons
Length of unit: 199ft 6in
Height: 12ft 8in
Width: 9ft 3in
Brake type: Auto air/electro-pneumatic
Nominal supply voltage: 25kV AC
Type of power collection: Overhead
Traction motor type: Metro-Vic
Available hp: 828
Number of traction motors: 4
Gear ratio: 70:17
Max unit speed: 75mph
Bogie type: Gresley design
Coupling type: Buck-eye
Coupling restriction: ScR EMU types
Formation and seating: DTS 83 second; MBS 70 second; BDTS 83 second

Below: *Following the Glasgow suburban electrification of the lines to Wemyss Bay and Gourock, in 1967 a batch of 19 4-car units of identical design to the Class 303 stock already in use was constructed by Cravens Ltd. In this illustration No 311095 stops at Bishopton on the 12.45 Gourock to Glasgow working during early 1980.* R. Ruffell

Above: *Marshalled between the DTS and BDTS coaches of Class 311 units is the power car, classified as MBS (Motor Brake Second). In this view taken from the brake van end, the Gresley design bogies can be seen. The roof mounted pantograph is in the usual position above the brake van. Underframe mounted boxes contain control and power equipment. MBS No Sc62181 of unit No 110 stands at Glasgow during early 1981.* K. Grafton

Below: *Most of the 311 units are now painted in standard blue and grey livery and many have the GG symbol painted on their body sides. This view of No 311104 near Drumry in 1979 shows the BDTS (Battery Driving Trailer Second) coach recognisable from the DTS (Driving Trailer Second) coach by having a mass of underslung equipment* T. Noble

Class 312

Unit numbers: 312201-312204
312701-312726
312781-312799
Introduced: 1975-1977
Built by: BREL York
Equipment manufacturer: GEC Traction Ltd
Area of use: LMR main line/GE, GN lines
Weight of unit: 151ton
Length of unit: 265ft $8\frac{1}{2}$in
Height: 13ft $0\frac{1}{2}$in
Width: 9ft 3in
Brake type: Auto air/electro-pneumatic
Nominal supply voltage: 25kV AC
Type of power collection: Overhead
Traction motor type: English Electric F46F
Available hp: 1,080
Number of traction motors: 4
Gear ratio: LMR units 63:17
ER units 61:19
Max unit speed: LMR units 75mph
ER units 90mph
Bogie type: BP14 and BT8
Coupling type: Buck-eye
Coupling restriction: Within ER/LMR type
Formation and seating: BDTS 84 second; MBS 68 second; TS 98 second; DTC 25 first/47 second

Below: *Following electrification of the Kings Cross outer suburban area at 25kv BRE York constructed the first batch of units under the classification of 312. 26 units were built and classified 312/0 and numbered 312001 to 312026 but were later renumbered in the 312701 to 312726 series. Units are of the same basic design as Class 310s working on the Midland line out of Euston, but front end redesigning under the windows gave an identification factor. In this view of unit No 312001 later 312701 the DTC (Driving Trailer Composite) is leading.* British Railways

Above: *Four units of the same class but classified 312/2 and numbered 312201-312204 were built in 1976 for use in the Birmingham area. These units are recognisable from the 310 counterparts in having flat front windows and a slightly modified front end layout under cab window height. No 312203 in this illustration stands at Rugby on 12 August 1976.* K. Lane

Below: *A further fleet of 19 4-car units of the same basic design were constructed for use on the GE lines out of Liverpool Street. These units are numbered in the series 312781 to 312799 and are all allocated to Clacton Shed. In this view of two units Nos 312785 and 312787 pass Stratford on a Liverpool Street to Southend Victoria working.* Colin J. Marsden

Above: *Front end equipment of Class 312/1 unit: (**1**) Multiple unit control jumper, (**2**) Main reservoir pipe, (**3**) Brake pipe, (**4**) Multiple unit control jumper socket, (**5**) Red rear tail light, (**6**) Buck-eye coupling, (**7**) Air warning horns, (**8**) AWS receiving apparatus. Unit No 312799 was photographed at Liverpool Street during late 1980. This front end layout is applicable to all Class 312 units.* Colin J. Marsden

Below: *This interesting photograph shows a Class 312/1 in original condition painted in all blue livery and carrying its original running number of 312101 with the new number affixed in the centre window. The use of headcodes on these units has now ceased and white spot markers will soon be applied. No 312101 in this photograph passes Stratford in East London.* Colin J. Marsden

Above: *MBS (Motor Brake Second) No E62523 of unit No 312794 shows the saddle boxes slung between the bogies containing all electrical equipment. These MBS cars seat 68 second class passengers compared to 70 second class passengers on the Class 310 units of the same basic design.* Colin J. Marsden

Below: *TS (Trailer Second) No E71207 of unit No 312794 shows the high density external body design with eight single leaf doors each side, with a total of 98 seats formed in the 3+2 format. Underframe equipment is sparce, with only brake equipment being carried.* Colin J. Marsden

Class 313

Unit numbers: 313001-313064
Introduced: 1976
Built by: BREL, York
Equipment manufacturer: GEC Traction Ltd
Area of use: GN inner suburban
Weight of unit: 103ton
Length of unit: 199ft 7in
Height: 13ft 0½in
Width: 9ft 3in
Brake type: Electro-pneumatic/dynamic
Nominal supply voltage: 25kV AC/750 DC
Type of power collection: Overhead/ third rail
Traction motor type: GEC G310AZ
Available hp: 880
Number of traction motors: 8
Gear ratio: 69:14
Max unit speed: 75mph*
Bogie type: BX 1
Coupling type: Tightlock
Coupling restriction: Within type only
Formation and seating: DMS 74 second; TS 84 second; BDMS 74 second

*30mph while operating from DC power supply.

Note: Some of these units have been reallocated to the GE Section and have had their third-rail equipment, dynamic brake systems, isolated.

Below: *The first production batch of 1972 designed new generation EMU stock emerged in 1976 when the first of 64 3-car units were introduced on the newly electrified Kings Cross/Moorgate electrification system. These 3-car sets have the unique feature of being able to work from 25kv AC overhead or 750v DC third rail with only the minimum of change over procedure. No 313021 arrives at Hatfield in this 1977 view.* Colin J. Marsden

Above: *This photograph of the original experimental high density stock used on the Southern Region during the early 1970s shows that only minor design changes were adopted in the production classes, namely the straightening of the sloping front, new design coupling equipment, and re-positioning of the air horns in the body shell. Two of the original PEP units stand outside the Railway Technical Centre at Derby during 1979.* Colin J. Marsden

Below: *Overhead power is collected via a roof mounted pantograph on the TS (Trailer Second) coach. Power is then transferred by means of power jumpers to the adjoining DMS (Driving Motor Second) and BDMS (Battery Driving Motor Second) cars and thence to the traction equipment. This view of No 313057 was taken near Hornsey during the spring of 1978.* L. Bertram

Above: *These units are immediately recognisable in the Kings Cross area as no other units of this design work in the locality. For explanation of front end equipment see Class 508 section. Here two units work in multiple on a Kings Cross to Welwyn Garden City local between Hornsey and Wood Green on 3 June 1979.* Brian Morrison

Below: *The prototype unit No 313001 in this view stands at Bayford on the Hertford Loop line during one of the commissioning trips. On early units the roof mounted air intakes for passenger ventilation were much larger than on later units as can be seen in this illustration.* British Railways

Below: *While working in the tunnel section of the line between Drayton Park and Moorgate power is collected via third rail collector shoes mounted on the inner bogie of the driving cars. Drayton Park is the change over point from 25kv AC to 750v DC and units must be brought to a stand at this station for the change over procedure to take place. All units are currently painted in blue and grey livery. In this illustration No 313021 departs from Old Street en route for Hertford East.* British Railways

Class 314

Unit numbers: 314201-314216
Introduced: 1979
Built by: BREL York
Equipment manufacturer: GEC Traction Ltd*
Area of use: Glasgow suburban
Weight of unit: 102ton
Length of unit: 199ft 7in
Height: 13ft 0½in
Width: 9ft 3in
Brake type: Electro-pneumatic/dynamic
Nominal supply voltage: 25kV AC
Type of power collection: Overhead
Traction motor type: GEC G310
Available hp: 880
Number of traction motors: 8
Gear ratio: 69:14
Max unit speed: 75mph
Bogie type: BX 1
Coupling type: Tightlock
Coupling restriction: Within type only
Formation and seating: DMS 68 second; TS 76 second; BDMS 68 second
*312201-312206 have Brush equipment.
Note: Units are fitted with thyristor control equipment

Below: *A further batch of 1972 design high density stock classified as 314 was introduced from 1979 for use on the Glasgow suburban system. Trains are formed of three cars and are allocated to Glasgow Hyndland Depot. In this view unit No 314214 stands at Partick station with a DMS (Driving Motor Second) leading.* J. N. Faulkner

Above: *Close up of 314 DMS (Driving Motor Second) coach showing the GG Trans-Clyde embellishment on the side rather than the more commonplace BR double arrow logo. Traction equipment and thyristor control equipment can be seen slung between the bogies. Guard controlled double leaf sliding doors are provided in passenger accommodation.* British Railways

Below: *Photographed on a rather cold and snowy day at Motherwell, TS (Trailer Second) No Sc71451 of unit No 314202 clearly shows the roof mounted pantograph lay out, equipment between the bogies consists of coach heating and control equipment* K. Grafton

Class 315

Unit numbers: 315801-315861
Introduced: 1980
Built by: BREL York
Equipment manufacturer: GEC Traction Ltd
Area of use: Liverpool Street-Shenfield
Weight of unit: 125ton
Length of unit: 264ft 10in
Height: 13ft $0\frac{1}{2}$in
Width: 9ft 3in
Brake type: Electro-pneumatic/dynamic
Nominal supply voltage: 25kV AC
Type of power collection: Overhead
Traction motor type: GEC G310
Available hp: 880
Number of traction motors: 8
Gear ratio: 69:14
Max unit speed: 75mph
Bogie type: BX 1
Coupling type: Tightlock
Coupling restriction: Within type only
Formation and seating: DMS 74 second; TS 84 second; TS 86 second; DMS 74 second

Left: *Designed as a replacement for GE Class 306 units, this batch of 61 4-car units of 1972 design started to appear in traffic late in 1980, all are allocated to Ilford Depot and carry numbers 315801-315861. Unit No 315820 travels south of Colchester with a driver training trip during March 1981.* M. Collins

Above: *It is interesting to note that although built during 1980 and 1981, this fleet of units has been fitted with the old fashioned four character headcode system, which is displayed above the road learner's window. Unit No 315811 with DMS No E64481 leading departs from Romford on a trial run during late 1980.* K. Grafton

Below: *Power is collected via a roof mounted single arm pantograph mounted on one of the TS vehicles. Associated electrical equipment is mounted in under slung boxes between the bogies. Passenger double leaf doors are passenger push button operated, the first of the type used on the Eastern Region. TS No E71401 of unit No 315813 was photographed at Stratford during November 1980.* Colin J. Marsden

Class 370 (APT)

Unit numbers: 370001-370006
Introduced: 1978-1979
Built by: BREL Derby
Equipment manufacturer (electrical): Asea
Area of use: Euston-Glasgow route
Weight of unit: 232ton
Length of unit: 482ft 0½in
Height: 11ft 6in
Width: 8ft 11in
Brake type: Air/hydraukanic
Nominal supply voltage: 25kV AC
Type of power collection: Overhead
Traction motor type: Asea LJMA410F
Available hp: 4,000
Number of traction motors: 4
Overall transmission ratio: Low speed 1.95739:1
High speed 1,7050:1
Max unit speed: 125mph
Bogie type: BT11, BT12, BT17
Coupling type: Bar
Coupling restriction: Within units only
Formation and seating: DTS 52 second; TS 72 second; TRSB 28 second; TU 43 unclassified; TF 47 first; TBF 25 first; M none

Above: *An APT power car, developing a total of 4,000hp stands outside Derby Works during 1977. Each power car is supported on two BP17a bogies fitted with ASEA LJMA410F traction motors. Power is collected via one single arm pantograph of the standard pattern.* British Railways

Below left: *Britain's most powerful electric multiple unit train is the APT or Class 370 unit. Trains are formed of seven coaches which form half of a service set. This view of No 370006 clearly shows the aerodynamic body design.* British Railways

Below: *The distinctive front end shape is apparent in this view of a DTS (Driving Trailer Second), outer ends of which are supported on a BT12 bogie. Behind the driver's cab is situated a small engine compartment which contains a diesel alternator set, hydrokinetic brake unit control equipment and a battery set.* British Railways

Below: *At the time of writing all coaches are allocated to Shields Road electric shed in Glasgow. Seating in second class cars is in the 2+2 formation, but is the more leisurely 2+1 for first class passengers. In the 'P' train the DTS (Driving Trailer Second) seats 52 passengers, DTS No Sc48102 stands outside Shields Road during 1979.* T. Noble

Above: *A test formation of APT stock runs near the Scottish borders on a trial run during May 1980. The grill cover on the cab nose end below the driver's window covers the two tone warning horns. It can be seen from this illustration that all underfloor equipment is covered in air smoothed casing which greatly reduces the aerodynamic stresses on the train.* British Railways

Above: *For emergency or hauling purposes, conventional coupling equipment is carried under the nose end as can be seen from this view of No 370006. The buffers are 'stored' in a vertical position and have to be turned through 90° prior to use. A conventional buck-eye coupling is 'stored' in the lowered position hanging on a drawhook, main reservoir pipes and a brake pipe are also supplied enabling the unit to be hauled by any air braked locomotive.* T. Noble

Below: *TS (Trailer Second), TRSB (Trailer Restaurant Second Buffet), TU (Trailer Unclassed) and TF (Trailer First) cars are mounted on BT11 trailer bogies carrying the cars in an articulated form. In this view of two cars above a BT11 bogie one of the large single leaf passenger sliding doors can be seen, which when shut lines up flush with the body side.* T. Noble

Class 405

Class type: 4SUB
Unit numbers: 4277-4754
Computer numbers: 405401 series
Introduced: 1948-1951
Built by: BR Eastleigh
Equipment manufacturer: GEC Traction Ltd
Area of use: Southern suburban lines
Weight of unit: 134ton
Length of unit: 257ft 5in
Height: 12ft 8$\frac{3}{4}$in
Width: 9ft 3in
Brake type: Auto air
Nominal supply voltage: 660-750V
Type of power collection: Third rail
Traction motor type: English Electric 507
Available hp: 1,000
Number of traction motors: 4
Gear ratio: 65:16
Max unit speed: 75mph
Bogie type: 43ft type
Coupling type: Screw
Coupling restriction: Within type
Formation and seating: DMBS 82 second; TS 120 second; TSO 120 second; DMBS 82 second. Seating may be different on certain stock

Right: *Front end detail of 4SUB (Class 405) unit: (**1**) Control jumper, (**2**) Power jumper socket, (**3**) Lighting control jumper socket, (**4**) Lighting control jumper, (**5**) Power jumper, (**6**) Control jumper socket, (**7**) Main reservoir pipe, (**8**) Brake pipe, (**9**) Draw hook, (**10**) Stencil headcode. In this view the driver fixes the headcode stencil on the front end, a task which no longer has to be done as all SR units now have roller blind headcode systems.* Colin J. Marsden

Below: *The 4SUB (Class 405) units are the oldest passenger carrying stock in use on the Southern Region main line, some cars being introduced from as early as 1946. Units are allocated to Wimbledon East and Selhurst Sheds and are all scheduled for early withdrawal. Unit No 4666 in this illustration departs from Waterloo on a Hampton Court service.* Colin J. Marsden

4299
1
2
3
4
5
6
7
8
9
10
MAY & PHILPOT

Above: *Units are formed of four coaches with a DMBS (Driving Motor Brake Second) at each end. All power and traction control equipment is carried between the bogies of these vehicles. DMBS (Driving Motor Brake Second) cars are of the same design at both ends and several units are in traffic with coaches from other unit formations. No 4627 approaches Wimbledon on a Waterloo to Chessington train during the summer of 1980.* Colin J. Marsden

Below: *Power at 750v DC is collected via an outside third rail pick up slipper mounted on the side of the power bogie. This view of the bogie of unit No 4629 shows the shoe (not in contact with the conductor rail), shoe beam, and shoe fuse box (under guard's door).* Colin J. Marsden

Above: *Between the DMBS cars are sandwiched TS (Trailer Second) and TSO (Trailer Second Open) coaches, TS coaches are built with 10 compartments each seating 12 passengers. This view of TS No S8935S of unit No 4655 shows the external layout and very empty underframe, housing only Westinghouse brake equipment (not visible in photograph). Note: All compartment cars were taken out of service during 1981 and replaced by TSO vehicles from withdrawn units.* Colin J. Marsden

Below: *4SUB units are incompatible with any other type of SR stock and therefore their workings are rather restricted. Units can usually be found working suburban services from London Waterloo and Victoria to many of the outer London branch lines such as Hampton Court, Shepperton and Beckenham Junction. Unit No 4662 stands at Selhurst during 1979.* Colin J. Marsden

Above: *Slight front end modifications exist between the early batches of 42XX units and the later 46XX and 47XX fleets. On 42XX units a grab rail is placed above the front cab window and step boards are fixed above the buffers. If units Nos 4732 and 4283 are compared this detail difference will become apparent. This line up of stock was photographed at Coulsdon North stabling point during 1979.* Colin J. Marsden

Below: *Several units have for some years carried a slightly modified front end, as roller blind route indicators have been fitted (of two different designs) and a modified window wiper unit which traverses the window horizontally rather than vertically as before. Unit No 4721 of SU (Selhurst) shed stands at Clapham Junction displaying this modification during December 1980.* Colin J. Marsden

Right: *This plate has been included to show the front end coupling of two 4SUB (Class 405 units). Jumper cables above buffer beam height consists of, power, control and, lighting. While those below buffer beam height are for brake pipe and, main reservoir connections, together with the screw coupling. In this illustration units Nos 4635 and 4619 stand at Clapham Junction station during early 1981.* K. Grafton

Class 410

Class type: 4BEP
Unit numbers: 7001-7022
Computer numbers: 410101 series
Introduced: 1957
Built by: BR Eastleigh
Equipment manufacturer: English Electric
Area of use: SR main lines
Weight of unit: 146ton
Length of unit: 265ft $8\frac{1}{2}$in
Height: 12ft 6in
Width: 9ft 3in
Brake type: Auto air/electro-pneumatic
Nominal supply voltage: 660-750V DC
Type of power collection: Third rail
Traction motor type: English Electric 507
Available hp: 1,000
Number of traction motors: 4
Gear ratio: 61:20
Max unit speed: 90mph
Bogie type: Mk 4/Commonwealth
Coupling type: Buck-eye
Coupling restriction: Within SR units
Formation and seating: DMBS 56 second; TC 24 first/24 second; TRB 21 unclassified; DMBS 56 second

Below: *Being introduced as part of the Kent Coast electrification scheme of the mid-1950s a fleet of 22 4-car buffet units were constructed being of identical design to 4CEP (Class 411 stock). Unit No 7012 in this illustration works an up train near Smarden during 1979.* Colin J. Marsden

Above: *Power is collected via a standard third rail collection shoe mounted on the outer end of each DMBS (Driving Motor Brake Second) car. Traction equipment is carried in Pannier boxes mounted between the bogies. Front end connections are of identical layout to other SR units of the same body appearance. Unit No 7020 stands at Windsor & Eton Riverside on a Royal train working.* J. Scrace

Below: *During the mid-1960s the TRB (Trailer Restaurant Buffet) No S69021 of unit No 7022 was selected for air conditioning experiments and this equipment is still currently fitted. The coach is recognisable as the quarter light window in the main panels are of a non-opening type. In this view No 7022 nears Smarden during the spring of 1978.* Colin J. Marsden

Right: *Front end detail of 410 and 411 stock:* ***(1)*** *Brake pipe,* ***(2)*** *Main reservoir pipe,* ***(3)*** *Multiple unit control jumper,* ***(4)*** *Multiple unit control jumper socket,* ***(5)*** *Power jumper socket (for shed supply),* ***(6)*** *Buck-eye coupling,* ***(7)*** *DSD isolation cock,* ***(8)*** *AWS isolation cock.*
Colin J. Marsden

Below: *4BEP (Class 410) units are recognisable from 4CEP (Class 411) units only by the omission of the TS (Trailer Second) coach and the fitting of a TRB (Trailer Restaurant Buffet) car which can seat 21 passengers. Units can usually be found marshalled with Class 411 stock on London to south-east coast services. All units are due to be refurbished during the period 1980-1983.* Colin. J. Marsden

411605
411605
1
2
3
5
2
1
4
411/2 4CEP
Restriction 4
7
6
8

Class 411

Class type: 4CEP
Unit numbers: 7101-7211
Computer numbers: 411501 series
Introduced: From 1957
Built by: BR Eastleigh
Equipment manufacturer: English Electric
Area of use: SR main lines
Weight of unit: 142ton
Length of unit: 265ft $8\frac{1}{2}$in
Height: 12ft 6in
Width: 9ft 3in
Brake type: Auto air/electro-pneumatic
Nominal supply voltage: 660-750V DC
Type of power collection: Third rail
Traction motor type: English Electric 507
Available hp: 1,000
Number of traction motors: 4
Gear ratio: 61:20
Max unit speed: 90mph
Bogie type: Mk 4 (some units have Mk 3 and Commonwealth bogies)
Coupling type: Buck-eye
Coupling restriction: Within SR units
Formation and seating: DMBS 56 second; TC 24 first/24 second; TS 64 second; DMBS 56 second

Right: *The first four units were constructed with the older system of contactor control for power take up, where later units were fitted with a camshaft control, a system which has been used on all subsequent SR EMU designs. The only other differences between the two batches were in internal decor, which used plastic and other ceramics in place of the wood veneer on earlier units. In this 1980 view No 7124 climbs away from Borough Green & Wrotham.* Colin J. Marsden

Below: *Designed as the base power for the Kent coast electrification a fleet of 110 4-car express geared units were constructed at Eastleigh Works from 1954 with the first unit entering service during 1955. In this view of one of the later units No 7202 climbs towards Sevenoaks tunnel during the summer of 1978.* Colin. J. Marsden

7124
7124
94

Above: *The DMBS (Driving Motor Brake Second) cars at each end are of identical design and behind the driver's cab is situated the guard's brake van with the motor bogie underneath which contains two 500hp English Electric traction motors. Control equipment is contained in underframe mounted panniers; unit No 7144 with DMBS No S61371 leading stands at Borough Green & Wrotham Station during 1979.* Colin J. Marsden

Below: *Unit Nos 7101 to 7152 are mounted on BR standard Mk 4 bogies. From No 7155 to No 7210 the TC (Trailer Composite), TS (Trailer Second) cars together with the inner ends of the DMBS (Driving Motor Brake Second) cars are mounted on BR commonwealth bogies. This photograph at Paddock Wood shows the commonwealth bogies and Mk 4 design under the same unit.* Colin J. Marsden

Above: *The TS (Trailer Second) vehicles of 4CEP (Class 411) units are of the open 2+2 layout, with a lavatory at one end. As can be seen from this photograph rain water from the roof water strip runs from the end of the coach to the ground, on some units a modified rain strip has been fitted to carry water to buffer beam height. No S70520 of unit No 7171 stands at Chatham during the summer of 1978.* Brian Morrison

Below: *Following a BRB decision in 1975 that no new main line electric multiple unit stock would be made available to the Southern for many years, a refurbishment programme of Class 411 stock was undertaken. Unit No 7153 was selected for experimental refurbishing and emerged from Eastleigh Works during late 1975. Major alterations were carried out, including moving the guard's van from DMBS vehicles and forming a new TBC (Trailer Brake Composite) coach in the middle of the formation. Unit No 7153 after being refurbished was photographed near Quarry tunnel during 1978.* B. Denton

Above: *Following evaluation of unit No 7153 a major refurbishing plan was authorised and BRE Swindon Works was given the contract to refurbish the complete fleet of 4CEP units. the first unit to appear was formerly No 7133 but carried its new computerised number of 411506. In this view the DMS (Driving Motor Second) No S61349 can be seen. During the refurbishment work all windows were replaced with a modern aluminium framed, tinted glass type, interiors were completely modernised with new seat designs, lower ceilings and fluorescent lighting. No 411506 stands at Waterloo on a commissioning trip on 18 December 1979.* Colin J. Marsden

Below: *TS (Trailer Second) vehicles were brought into line with modern standards, with new seating, lighting and decoration. The refurbished units are easily recognisable from original units by the modern window design and coach layout. Most refurbished units are mounted on commonwealth bogies (unpowered) and B4 bogies (powered).* Colin J. Marsden

Above: *During the refurbishing the old TC (Trailer Composite) coaches were completely rebuilt into a new type of car not previously used on the Southern. The coaches are classified TBC (Trailer Brake Composite) and are formed (from left to right of illustration) toilet, second class compartment seating six, brake van, guard's office, four first class compartments seating 24 and another toilet.* Colin J. Marsden

Class 411 (Refurbished)

Class type: 4CEP
Unit numbers: 411501 series
Refurbished: From 1980
Refurbished by: BREL Swindon
Equipment manufacturer: GEC Traction Ltd
Area of use: SR main lines
Weight of unit: 156ton
Length of unit: 265ft 8½in
Height: 12ft 6in
Width: 9ft 3in
Brake type: Auto air/electro-pneumatic
Nominal supply voltage: 660-750V DC
Type of power collection: Third rail
Traction motor type: English Electric 507
Available hp: 1,000
Number of traction motors: 4
Gear ratio: 61:20
Max unit speed: 90mph
Bogie type: Mk 4 and Commonwealth
Coupling type: Buck-eye
Coupling restriction: Within SR units
Formation and seating: DMS 64 second; TBC 24 first/6 second; TS 64 second; DMS 64 second

Class 414

Class type: 2HAP
Unit numbers: 6001-6173
Computer numbers: 414001-414173
Introduced: From 1957
Built by: BR Eastleigh
Equipment manufacturer: English Electric
Area of use: SR outer suburban and main lines
Weight of unit: 70ton
Length of unit: 132ft 8½in
Height: 12ft 9½in
Width: 9ft 3in
Brake type: Auto air/electro-pneumatic
Nominal supply voltage: 660-750V DC
Type of power collection: Third rail
Traction motor type: English Electric 507
Available hp: 500
Number of traction motors: 2
Gear ratio: 61:20
Max unit speed: 90mph
Bogie type: Mk 4/Commonwealth
Coupling type: Buck eye
Coupling restriction: Within SR units
Formation and seating: DMBS 84 second; DTC 19 first/50 second

Below: *Fortytwo 2-car units were built from 1957 at Eastleigh and classified 2HAP, these units were built as a replacement for the ageing 2HAL units on the London to Maidstone line. Units were fitted with 1951 type contactor control equipment and are known within railway circles as 51 stock. Here No 6023 of the original batch approaches Lewes during 1979.* Colin J. Marsden

Above: *Only one of the two cars is powered, the DMBS (Driving Motor Brake Second).This coach is alway recognisable from the DTC (Driving Trailer Composite) by a black triangle painted on the front end below the headcode box, denoting to operating staff that power equipment and guard's van are situated at the end of the unit. Unit No 6013 stands at East Wimbledon during 1981.* Colin J. Marsden

Below: *A further batch of 63 units were constructed late in 1958 and were fitted with a new 'camshaft' type control system. These became known to operating staff as 57 stock. Units are recognisable from the earlier batch by roof detail. All cars of the original batch were fitted with roof mounted conduit where later units in the number series 6043 to 6105 are only fitted with conduit on the DTC coach. Unit No 6088 shows the roof conduit on one coach only when photographed during 1978.* Colin J. Marsden

6143
96

Private

Above left: *The next batch of units to appear, Nos 6106 to 6146, were again of slightly different design, with Mk 3B bogies fitted to DMBS cars and commonwealth to DTC cars. Again roof detail identifies this batch from others of the same design, by being clear of all conduit with only ventilators mounted on the roof. Unit No 6143 illustrated here demonstrates this point.* Colin J. Marsden

Left: *Several different designs of rain water drainage system have been tested on 2HAP units. Unit No 6107 on the right of this photograph is fitted with an additional down pipe in front of the driver's side window.* Brian Morrison

Above: *The final batch of units Nos 6147 to 6173 was built in 1963 and are of yet another design with slightly smaller headcode box and roof mounted angled rain water strips. Power bogies of this batch reverted to the originally used Mk 4 type. Unit No 6158 passes Headcorn during 1980.* Colin J. Marsden

Below: *All power and control equipment is housed in underframe mounted boxes under the DMBS (Driving Motor Brake Second) coach. The only underframe equipment carried on DTC (Driving Trailer Composite) cars is brake reservoir and EP equipment. Unit No 6165 stands at Paddock Wood on a Maidstone service.* Colin J. Marsden

Class 415

Class numbers: 415/1, 415/2, 415/4
Class type: 4EPB
Unit numbers: 5001-5264 (415/1)
5301-5370 (415/2)
5401 fleet (415/4)
Computer numbers:
415001-415264 (415/1)
415301-415370 (415/2)
415401 fleet (415/4)
Introduced: From 1951 (415/1)
From 1957 (415/2)
From 1979 (415/4)
Built by: BR Eastleigh
Equipment manufacturer: GEC Traction Ltd
Area of use: SR inner/outer suburban
Weight of unit: 136ton
Length of unit: 275ft 5in (415/1, 415/4)
263ft $11\frac{1}{4}$in (415/2)
Height: 12ft $9\frac{1}{2}$in
Width: 9ft 3in
Brake type: Auto air/electro-pneumatic
Nominal supply voltage: 660-750V DC
Type of power collection: Third rail
Traction motor type: EE 507
Available hp: 1,000
Number of traction motors: 4
Gear ratio: 65:16
61:19 (some units of Class 415/2)
Max unit speed: 75mph
90mph (some units of Class 415/2)
Bogie type: 40ft type
Coupling type: Buck-eye
Coupling restriction: Within SR Units
Formation and seating: MBSO 82 second; TS 120 second*; TSO 112 second; MBSO 82 second (415/1, 415/4)
MBSO 84 second; TS 112 second; TS 112 second; MBSO 84 second (415/2)
*Some cars seat only 108 in a few Class 415/1 units.

Note: Unit No 5246 and 5262 are formed of three 415/1 cars and one 415/2 MBSO, giving total of 388 seats. Unit No 5261 is formed of 415/2 MBSO cars and 415/1 TS coaches, and seats a total of 408.
Unit Nos 5301-5302 are formed of 415/2 MBSO cars and 415/1 TS vehicles, and seat 390.

Below: *A more modern version of the 4SUB (Class 405) units emerged in 1951, when the first of a fleet of 4-car suburban units of Southern Region design entered service fitted with a new electro pneumatic brake system. Also used for the first time on EMU stock was the buck-eye coupling and rubbing plate. Units were numbered in the 5001 range ceasing at 5053 and recommencing with No 5101 and finally finished at No 5260, a total of 213 units. In this view unit No 5110 passes one of the earlier 4SUB units at Wimbledon during the summer of 1980. Units of this type are known to operating staff as 51 stock and are classified 415/1* Colin J. Marsden

Above: *4EPB Class 415/1 front end detail: (**1**) Brake pipe, (**2**) Main reservoir pipe, (**3**) Multiple unit control jumper (27 way), (**4**) Power jumper socket (for shed use only), (**5**) Multiple unit control jumper socket, (**6**) Rubbing plate, (**7**) Buck-eye, (**8**) Buck-eye release chain. Unit No 5122 was photographed during early 1980* Colin J. Marsden

Above: *Unit Nos 5156 to 5260 are recognisable from their earlier classmates in having an additional rain strip just above door height. This alteration can clearly be seen if this and the plate of No 5110 on the previous page are compared. Unit No 5184 in ex-works condition poses for the camera at East Wimbledon early in 1980.* Colin J. Marsden

Below: *The two TS (Trailer Second) cars are normally of a different internal design. One vehicle is usually a 10 compartment coach seating 12 people per bay, while the other vehicle is of the saloon layout and only seats a total of 102 passengers. Due to some reforming of units following collision or other damage some trains may be in service formed with two identical TS vehicles. The only underframe mounted equipment carried on TS vehicles is for brake systems. Unit No 5126 approaches Fulwell junction with a Waterloo service during the summer of 1980. Note: As this book goes to press several units were being reformed to give all open accommodation.* Colin J. Marsden

Above & below: *During the mid-1970s four of the 2-car SAP units of Class 418 were disbanded and gave their DMBS (Driving Motor Brake Second) cars to new 4-car EPB units; trailer cars were either rebuilt from class 405 types or acquired from withdrawn units of Class 415. Unit numbers allocated were No 5263 onwards. Units No 5263 was refurbished and fitted with new upholstery, modern fluorescent lighting and painted in blue and grey Inter City livery. Other units soon followed by being refurbished at various works. Unit No 5263 in the upper illustration departs from Waterloo during 1978, while in the lower illustration refurbished No 5225 stands at Strawberry Hill Depot. An interesting feature is that this example has its numbers above the cab window rather than above the headcode box. At the present time refurbished units are being renumbered in the 54XX range and classified as 415/4s.*
Both Colin J. Marsden

Above & below: *A further batch of 4EPB units was constructed during 1960, but this time to BR coach designs and fitted with 1957 camshaft control equipment, numbering was in the Nos 5301 to 5370 range and are now classified 415/2. Units are recognisable from the SR designed units by the following points: flat nose ends and body sides, roofs are clear of all electrical conduit and body side windows are of a different design, with no top quarter light in the door panel. If the upper plate of No 5340 is compared with unit No 5263 on the previous page the differences will be seen. The lower plate of unit No 5317 shows the plain roof design with only ventilators protruding.* Colin J. Marsden/Brian Morrison

Below: *Two interesting units are those numbered 5301 and 5302. These are formed of BR standard design DMBS (Driving Motor Brake Second) coaches and SR design TS (Trailer Second) coaches. These two units are easily recognisable from side or roof detail. This view of No 5301 at Higham clearly shows the two different roof designs.* Colin J. Marsden

Class 416

Class numbers: 416/1, 416/2
Class type: 2EPB
Unit numbers: 5651-5684 (416/1)
5702-5795 (416/2)
Computer numbers:
416651-416664 (416/1)
416702-416799 (416/2)
Introduced: 1953 (416/1)
1953-1956 (416/2)
Built by: BR Eastleigh
Equipment manufacturer: GEC Traction Ltd
Area of use: SR inner/outer suburban
Weight of unit: 70ton (416/1)
71ton (416/2)
Length of unit: 129ft 6½in (416/1)
132ft 8½in (416/2)
Height: 12ft 9½in
Width: 9ft 3in
Brake type: Auto air/electro-pneumatic
Nominal supply voltage: 660-750V DC
Type of power collection: Third rail
Traction motor type: EE507
Available hp: 500
Number of traction motors: 2
Gear ratio: 65:16
Max unit speed: 75mph
Bogie type: 40ft type (416/1)
Mk 3D (416/2)
Coupling type: Buck-eye
Coupling restriction: Within SR units
Formation and seating: DMBS 84 second; DTS 94 second (416/1)
DMBS 84 second; DTS 102 second (416/2)

Below: *Three different fleets of units come under the classification of 416 or 2EPB. The first are 34 SR built units in the number range 5651 to 5684 and classified 416/1. Built in 1958 to replace aged units on the Waterloo Windsor and Weybridge line, these units hold the distinction of being the first to be introduced formed of an all saloon seating layout, units now work on the central and south western divisions of the Southern. Unit No 5652 approaches Chipstead on a Tattenham Corner service during 1979.*
Colin J. Marsden

Above: *DMBS (Driving Motor Brake Second) coaches are virtually identical to those in 4EPB (Class 415) units except that the black triangle sign is painted on the front end. DTS (Driving Trailer Second) cars are basically a TS (Trailer Second) coach with a driving end. All equipment is contained in the DMBS car and only the braking system is mounted under the DTS. Unit No 5667 with DTS No S16117 leading approaches Kingswood station during 1979.* Colin J. Marsden

Below: *In common with many other unit types on the Southern blue and grey inter-city colours have been applied from the late 1970s. Power is collected from both ends of the unit and is tranferred to the power equipment by means of a power train line, the fuse of which can be seen under the third passenger door from the front of unit No 5657 at Norwood Junction.* B. Denton

Above: *The second batch of 2EPB (Class 416 units) was built from 1954 and numbered in the series 5701 to 5779, these units are the BR version of the 416/1 and are classified 416/2. Body design is the typical flat fronted and slab sided design and are easily recognisable from the earlier batch. Units work on all three divisions of the Southern and can be seen working many inner and outer suburban duties. Unit No 5756 arrives at Waterloo coupled to one of the earlier 416/1 units during 1978.*
Colin J. Marsden

Below: *This photograph serves as a good comparison of the difference between the BR designed 2EPB (Class 416) and the SR designed 4EPB (Class 415). Front end connections and accessories are identical but the styling can be seen to be quite different. Unit Nos 5761 and 5029 stand at Stewarts Lane Depot during 1980.*
Colin J. Marsden

Above & Below: *The third and final batch of Class 416 units Nos 5781 to 5793 is also classified as 416/2 but differ in appearance to the main batch in having larger brake vans (1/3 of the DMBS car) and a smaller route indicator panel. These 15 units were originally built for and used on the Tyneside electrified system, but were transferred to the Southern during 1963. The upper plate shows No 5784 near Clapham Junction during 1979. In the lower plate unit No 5790 with DTS leading stands at Selhurst. From this view the empty underframe of the DTS car can clearly be seen*
Colin J. Marsden/B. Denton

Class 418

Class type: 2SAP
Unit numbers: 5604-5635
Computer numbers: 418604-418635
Introduced: 1957, converted during 1976*
Built by: BR Eastleigh
Equipment manufacturer: English Electric
Area of use: SR inner/outer suburban
Weight of unit: 72ton
Length of unit: 129ft 6in
Height: 12ft $9\frac{1}{2}$in
Width: 9ft 3in
Brake type: Auto air/electro-pneumatic
Nominal supply voltage: 660-750V DC
Type of power collection: Third rail
Traction motor type: English Electric 507
Available hp: 500
Number of traction motors: 2
Gear ratio: 61:20
Max unit speed: 75mph
Bogie type: 40ft type
Coupling type: Buck-eye
Coupling restriction: Within SR units
Formation and seating: DMBS 84 second; DTS 62 second

*Adapted for all second class operation during 1976.

Left: *The last of the Southern designed units to appear did not do so until 1957 when a fleet of 36 2-car units emerged from Eastleigh classified as 2HAP and numbered in the 5601 to 5636 range. Units were formed DMBS (Driving Motor Brake Second) and DTC (Driving Trailer Composite). During 1976 DTC cars were converted to second class only use and reclassified DTS (Driving Trailer Second) units were then designated 2SAP and classified 418. Unit No 5612 in this plate was photographed near Wimbledon during 1979.* Colin J. Marsden

Above: *DMBS (Driving Motor Brake Second) No S14554 leads in this view of No 5634 arriving at Clapham Junction, DMBS cars are identical to those of 4EPB (Class 415) units except that the black triangle sign is painted on the front end. All 2SAP (Class 418) units are currently allocated to Selhurst Shed and work on a multitude of Central Division suburban services.* K. Grafton

Below: *Due to accident damage unit No 5624 has been reformed with the spare DMBS (Driving Motor Brake Second) of 2HAP No 6017, therefore giving a one SR and one BR car formation. While formed in this way No 5624 is immediately recognisable. In this illustration No 5624 stands at Epsom Downs.* B. Denton

Class 419

Class type: MLV
Unit numbers: 68001-68010
Computer numbers: 419001-419010
Introduced: 1959
Built by: BR Eastleigh
Equipment manufacturer: English Electric
Area of use: SR South Eastern Division boat trains
Weight of unit: 45ton
Length of unit: 67ft 1in
Height: 12ft 9½in
Width: 9ft 3in
Brake type: Auto air/electro-pneumatic
Nominal supply voltage: 660-750V DC/traction battery
Type of power collection: Third rail
Traction motor type: English Electric 507
Available hp: 500
Number of traction motors: 2
Gear ratio: 61:20
Max unit speed: 90mph
Bogie type: Mk 3B
Coupling type: Buck-eye
Coupling restriction: Within SR units/ and vacuum stock
Note: These vehicles are fitted with a vacuum exhauster.

Below: *A unique fleet of 10 single car vehicles known as MLV (Motor Luggage Vans) were constructed at Eastleigh during 1959, each single car is equipped with a driving cab at each end and 500hp traction motors. MLVs can run singularly or in a formation of EMU stock, usually MLVs are to be found on the London end of Victoria to Dover boat trains. MLV No 68006 stands at Grosvenor Carriage Sidings in this view taken during 1979.* Colin J. Marsden

Above & below: *MLVs are an interesting breed as they are fitted with a vacuum exhauster, and traction battery set. The vacuum exhauster is utilised when hauling vacuum brake stock but due to traction limitations, are limited to a trailing load of 100 tons. The traction battery is utilised if the MLV and a load of up to 100 tons are required to work into sidings not equipped with third rail equipment. The driver after bringing his train to a stand operates a push button in the cab which lifts the shoes clear of the rail and changes power over to battery conditions. MLVs are restricted in the time that they are allowed to work on battery conditions and have to return to the third rail after about 20 minutes battery operation. Vehicles are painted in standard blue and grey livery and are all allocated to Ramsgate Shed. In the upper view No 68004 passes near Paddock Wood with an up boat train, while in the lower view No 68002 passes through Headcorn station.* Both Colin J. Marsden

Class 420

Class type: 4BIG
Unit numbers: 7031-7058*
Computer numbers: 420231-420258
Introduced: From 1964
Built by: BR York
Equipment manufacturer: English Electric, GEC Traction
Area of use: SR main lines
Weight of unit: 151ton
Length of unit: 265ft 8½in
Height: 12ft 9½in
Width: 9ft 3in
Brake type: Auto air/electro-pneumatic
Nominal supply voltage: 660-750V DC
Type of power collection: Third rail
Traction motor type: English Electric 507/16F
Available hp: 1,000
Number of traction motors: 4
Gear ratio: 61:20
Max unit speed: 90mph
Bogie type: Mk 3, B5
Coupling type: Buck-eye
Coupling restriction: Within SR units
Formation and seating: DTC 18 first/36 second; MBS 56 second; TRBS 40 second; DTC 24 first/28 second
*Unit Nos 7031-7048 fitted with electric parking brake.

Below: *4BIG (Class 420) units were first introduced in 1965 when 18 4-car buffet units were constructed at York for use on the Victoria to Brighton and South Coast line. The power car or MBS (Motor Brake Second) is situated in the middle of the formation, power is collected via shoes on the outer ends of the DTC (Driving Trailer Composite) cars. DTC cars are of different internal designs at each end of the unit, with one end having four first class compartments and the other having three first class compartments and one second. A further batch of 10 units of the same basic design was constructed from 1970 for use on the Waterloo-Portsmouth line. In this illustration No 7055 passes Clapham Junction during late 1980.* Colin J. Marsden

ıbove: *The buffet cars of 4BIG (Class 420) units are all of the same basic design, but mall differences do exist between them. On the earlier units (Nos 7031-7048) the ıternal layout was based on that of the older 4BEP (Class 410) units, with wood eneer being widely used. On latter units (Nos 7049-7058) plastic and ceramics were videly used. TRBS (Trailer Restaurant Buffet Second) No S69314 of unit No 7044 tands at Victoria during early 1981.* K. Grafton

elow: *This plate has been included to show the original livery applied to the first batch f 4BIG and 4CIG units in the early 1960s. The units were painted in standard EMU reen with small yellow warning panel painted in the vestibule door position, unit and oach numbers were painted in a gold/yellow colour. Early members of BIG/CIG units ere fitted with roof board clips, but few have ever been used.* John Vaughan

Class 421

Class type: 4CIG
Unit numbers: 7301-7438*
Computer numbers: 421701-421838
Introduced: 1964 and 1970
Built by: BR/BREL York
Equipment manufacturer: English Electric, GEC Traction
Area of use: SR main lines
Weight of unit: 148ton
Length of unit: 265ft 8½in
Height: 12ft 9½in
Width: 9ft 3in
Brake type: Auto air/electro-pneumatic
Nominal supply voltage: 660-750V DC
Type of power collection: Third rail
Traction motor type: English Electric 507/16F
Available hp: 1,000
Number of traction motors: 4
Gear ratio: 61:20
Max unit speed: 90mph
Bogie type: Mk 4, Mk 6 and B5
Coupling type: Buck-eye
Coupling restriction: Within SR units
Formation and seating: DTC 18 first/36 second; MBS 56 second; TS 72 second; DTC 24 first/28 second

*Units 7301-7336 fitted with electric parking brake.

Left: *The first of the main line replacement stock emerged from York during 1964 and was classified as 4CIG (now Class 421/1). The units were a break from the Southern tradition of having motors under the driving cars as all power equipment is housed in a MBS (Motor Brake Second) coach in the middle of the unit. The first batch to enter service Nos 7301-7336 was used on the Brighton line in company with the 4BIG (Class 420 units). 7307 of the original batch stands at Brighton during July 1968. Underframe mounted equipment is sparse on DTC (Driving Trailer Composite) cars with only toilet water and auxiliary reservoirs being housed.* British Railways

Above: *The second batch of 4CIG units were of the same external design and numbered 7337 to 7366 being classified as 421/2. These units were built for use on the Waterloo-Portsmouth line to replace the old 4COR series of units. The interior layout of this batch is of more modern design and the use of plastic, aluminium and ceramics replaced the wood veneer of the original batch. 4CIG No 7354 of the second batch leads a Waterloo to Portsmouth train near New Malden on 28 August 1980.* Colin J. Marsden

Below: *The third batch of units were of identical design to the previous fleet and introduced to replace ageing COR units on Waterloo, Reading and Guildford lines. This batch of units was introduced from 1971 and allocated running numbers 7367 to 7437. By the mid-1970s several units had been reallocated away from the south western division to Brighton for south coast working. In this illustration No 7369 pulls into Lewes with a Victoria service during 1979.* Colin J. Marsden

Above: *Some external differences exist on the MBS (Motor Brake Second) vehicles, on the first batch Nos 7301 to 7336 MBS cars are mounted on Mk 4 power bogies fitted with an electric parking brake, these bogies are recognisable as they resemble a Gresley designed bogie. All subsequent batches have been fitted with Mk 6 power bogies as illustrated in this plate of MBS No S62302 of unit No 7352. All traction and main control equipment is housed between the bogies.* Colin J. Marsden

Below: *This view shows the external appearance of the TS (Trailer Second) coach mounted on B5 bogies, these units have the low density seating layout of 2+2, with only six doors per car. Underframe equipment is confined to brake control units.* Colin J. Marsden

Above: *DTC (Driving Trailer Composite) cars are identical to those of 4BIG (Class 420) stock and the reference to first class accommodation in that section also applies. 4CIG (Class 421) units are immediately recognisable from 4VEP (Class 423) units described in the next section. On CIG units there are far less entrance doors on all vehicles, unlike the VEP units that have access doors to each seating bay. First class accommodation on CIG stock is at the outer end of the DTC car while on VEP stock first class accommodation is at the rear end of the DTC and the guard's brake van on CIG stock is of a smaller type than on VEPs. If this illustration of 4CIG No 7381 at Woking is compared with a plate in the VEP section these differences will be readily seen.*
Colin. J. Marsden

Below: *The last batch of CIG stock ran to only one unit, being allocated running number 7438. The unit was put into service during 1972 as a replacement for 4CEP (Class 411) No 7181 which had previously been withdrawn due to collision damage. No 7438 is of identical appearance to other members of Class 421/2, in this illustration the unit approaches Horsham during 1979.* Colin J. Marsden

Class 423

Class type: 4VEP
Unit numbers: 7701-7894
Computer numbers: 423701-423894
Introduced: From 1967
Built by: BR/BREL York
Equipment manufacturer: English Electric
Area of use: SR main lines
Weight of unit: 149ton
Length of unit: 265ft $8\frac{1}{2}$in
Height: 12ft $9\frac{1}{2}$in
Width: 9ft 3in
Brake type: Auto air/electro-pneumatic
Nominal supply voltage: 660-750V DC
Type of power collection: Third rail
Traction motor type: English Electric 507/16F
Available hp: 1,000
Number of traction motors: 4
Gear ratio: 61:20
Max unit speed: 90mph
Bogie type: Mk 6 and B5
Coupling type: Buck-eye
Coupling restriction: Within SR units
Formation and seating: DTC 24 first/38 second; TS 98 second; MBS 58 second; DTC 24 first/38 second

Below: *A total of 194 4-car high density units started to enter service from early 1967, the first batch of 20 units was destined for the newly electrified Waterloo-Bournemouth line. Numbering was in the range 7701 to 7720 and classification given was 4VEP. Coaches are of the same basic design as 4CIG units but are of the high density seating style (3+2). In this view No 7712 departs from Waterloo during 1979.* Colin J. Marsden

Above: *The next batch to be built Nos 7721 to 7755 was introduced from late 1967 and was destined for the Brighton line. Units are of identical appearance to the earlier batch. Units Nos 7701 to 7755 are recognisable from later units as they are fitted with lifting lugs that protrude at bogie centre position from the body side. Unit No 7746 showing this feature stands at Wimbledon East Depot during 1980.* Colin J. Marsden

Below: *The high-density body styling is clearly visible in this view of TS (Trailer Second) No S70935 of unit No 7784, seating in these cars is in the 3+2 style, giving a total of 98 second class seats. Underframe mounted equipment consists only of EP brake control equipments.* K. Grafton

Above: *This view of 4VEP (Class 423) MBS (Motor Brake Second) No S62221 of unit No 7760 shows the external body appearance, with the guard's van being nearer the camera. As will be seen the brake van size is far larger than on the 4CIG units. Underframe mounted boxes between the bogies, contain power and control equipment. Power bogies under the MBS coach are of the Mk 6 type.* K. Grafton

Below: *Soon further batches of VEP units were built to replace several old and worn out classes. After No 7755 all units are of identical design, and have no recognition features between them. Roof equipment is confined to coach ventilators. Unit No 7847 of a later batch delivered during 1973 passes Wallers Ash, Near Winchester during 1979.* Colin J. Marsden

Above: *Unit No 7755 and 7756 have been made recognisable from other units over the past couple of years, the two units were selected for window experiments, which involved the removal of the top opening windows in the large panels and replacing them with one large glass window plate. Unit No 7755 fitted with this modification passes Wimbledon during 1980.* Colin J. Marsden

Below: *This view shows the differences between 4CIG (Class 421) and 4VEP (Class 423) units. Front ends, nose ends, cabs and bogies are of identical design with only passenger accommodation being different. Up until 1979 4VEP units were fitted with orange or turquoise curtains, but due to financial cuts imposed these have now all been removed. 4CIG No 7376 and 4VEP No 7758 stand at New Cross Gate sidings during 1979.* Colin J. Marsden

Above: *4VEP (Class 423) front end layout:* ***(1)*** *27 way control jumper,* ***(2)*** *Main reservoir pipe,* ***(3)*** *Brake pipe,* ***(4)*** *27 way control jumper socket,* ***(5)*** *Buck-eye coupling,* ***(6)*** *Buck-eye release chain,* ***(7)*** *Driver's safety device isolating cock. 4VEP (Class 423) No 7812 stands inside Wimbledon Park Shed during 1980. This front end layout also applies to 4BIG, 4CIG and 4VEG units.* Colin J. Marsden

Above: *The 4VEP units are currently allocated to all three divisions of the Southern and can be seen working on most outer suburban and main line routes. In 1982 the following depots have an allocation of units: BM (Bournemouth), BI (Brighton), WD (Wimbledon) RE (Ramsgate) and FR (Fratton). Unit No 7821 passes Clapham Junction late in 1980 on a Waterloo-Guildford working.* Colin J. Marsden

Below: *When parked together the 4VEP (Class 423) looks quite modern next to the 1951-designed 4CEP (Class 411). Front end equipment is the same on both types but looks different on the VEP as pipes and connections are set in the front end rather than affixed to the front, as on CEP/BEP/EPB style stock. Photograph taken at Waterloo during late 1979.* Colin J. Marsden

Class 427

Class type: 4VEG
Unit numbers: 7901-7912
Computer numbers: 427901-427912
Converted: 1978 (converted from 4VEP Class 423 stock)
Built by: BR York
Equipment manufacturer: English Electric
Area of use: Victoria-Gatwick line
Weight of unit: 149ton
Length of unit: 265ft 8½in
Height: 12ft 9½in
Width: 9ft 3in
Brake type: Auto air/electro-pneumatic
Nominal supply voltage: 660-750V
Type of power collection: Third rail
Traction motor type: English Electric 507/16F
Available hp: 1,000
Number of traction motors: 4
Gear ratio: 61:20
Max unit speed: 90mph
Bogie type: Mk 6 and B5
Coupling type: Buck-eye
Coupling restriction: Within SR units
Formation and seating: DTC 24 first/34 second; TS 90 second; MBS 50 second; DTC 24 first/34 second

Below: *From the mid-1970s the need for special units to work the Gatwick Airport to Victoria rail link could be seen, and during early 1978 12 4VEP units (Nos 7788-7799) were converted to accommodate luggage in saloon mounted luggage racks, in place of some seating. After units were converted they were reclassified as 4VEG (Vestibule Electro Gatwick) and classified Class 427, being numbered 7901-7912. Unit No 7911 passes Tinsley Green near Gatwick during 1978.* Colin J. Marsden

Above: *After entering service units were endorsed with the 'Rapid City Link Gatwick-London' legend on the sides of the DTC (Driver Trailer Composite) coaches, also alongside the BR logo the aircraft sign was painted, making the units recognisable from 4VEP (Class 423). Unit 4VEG No 7904 stands off its correct route at Fratton.*
J. Scrace

Below: *DTC (Driving Trailer Composite) No S76521 of unit No 7909 clearly shows the side embellishment and repositioning of car number on the body side. The first window behind the driver's cab door is blanked out, as this is one of the set of seats that have been removed and luggage racks provided.* Colin J. Marsden

Class 430

Class type: 4REP
Unit numbers: 3001-3015
Computer numbers: 430301-430315
Introduced: 1966 and 1974
Built by: BR York
Equipment manufacturer: English Electric, GEC Traction
Area of use: Waterloo-Bournemouth line
Weight of unit: 173ton
Length of unit: 265ft $5\frac{1}{4}$in
Height: 12ft $9\frac{1}{2}$in
Width: 9ft 3in
Brake type: Auto air/electro-pneumatic
Nominal supply voltage: 660-750V DC
Type of power collection: Third rail
Traction motor type: English Electric 546/B
Available hp: 3,200
Number of traction motors: 8
Gear ratio: 61:19
Max unit speed: 90mph
Bogie type: Mk 6 and B5
Coupling type: Buck-eye
Coupling restriction: See note at front
Formation and seating: DMS 64 second; TRB 23 unclassified; TBF 24 first; DMS 64 second

Below: *When introduced to the Southern in 1967 these 4REP (Class 430) units were the most powerful electric multiple unit stock ever built developing a total of 3,200hp (This distinction has since been lost to Class 370 APT units). The REP units are a unique breed in many respects as each DMS (Driving Motor Second) coach is mounted on two Mk 6 power bogies each fitted with two powerful 400hp motors. Between the power bogies is a mass of power and control equipment. Unit No 3006 approaches Southampton during 1980.* Colin J. Marsden

Above: *A break from SR tradition was made with these units with the guards van being in a non-powered coach. Front ends are basically the same as the VEP/CIG style but can be recognised by a train heating cable hanging under the buffer on the driver's side. REP units are all allocated to Bournemouth Depot and work with TC stock on Waterloo-Bournemouth fast and semi-fast services. Unit No 3007 passes Hampton Court junction during late 1979 with an up service. It is normal practice for the REP set to be marshalled at the London end of the train.* Colin J. Marsden

Below: *All trailer cars for the REP units were rebuilt from existing loco hauled vehicles, rebuilding was carried out at York works. This work included rebuilding the coaches from frames upwards, fitting EP brakes, and EMU control equipment, modern upholstery and new bogies. In this view of the TRB (Trailer Restaurant Buffet) No S69320 of unit No 3002 the original livery of all blue is displayed, between the bogies a motor generator set, water tanks, EP brake control equipment and battery box will be noted.* British Railways

Class 485, 486

Class type: 4VEC, 3TIS
Unit numbers: 031-035 (486), 041-046 (485)
Computer numbers: 486031-486035 (486), 485041-485046 (485)
Introduced: 1967 (on SR)
Built by: Metropolitan Cammell, Union Car Co and Cammell Laird
Equipment manufacturer: BTH
Area of use: Isle of Wight
Weight of unit: 94ton (485), 65ton (486)
Length of unit: 207ft 11in (485), 156ft $5\frac{1}{2}$in (486)
Height: 9ft 6in
Width: 8ft 8in
Brake type: Auto air/electro-pneumatic
Nominal supply voltage: 630V DC
Type of power collection: Third rail
Traction motor type: English Electric
Available hp: 960 (485), 480 (486)
Number of traction motors: 4 (485) 2 (486)
Gear ratio: 63:17
Max unit speed: 45mph
Bogie type: LT Design type Z and V
Coupling type: Ward
Coupling restriction: Within type
Formation and seating: MBSO 26 second; TSO 42 second; DTSO 38 second; MBSO 26 second (485) MBSO 26 second; TSO 42 second; DTSO 38 second

Right: *Front end detail of unit No 031. All Isle of Wight stock is of the same basic design retaining the unique London Transport coupling system. It is normal operating practice to run trains formed of one VEC and one TIS unit, but during the summer months some single unit running occurs, particularly between Ryde Pier Head and Ryde Esplanade stations.* R. I. Wallace

Below: *Following the decision to withdraw steam traction from the Southern Isle of Wight railway system in the mid 1960s — BR purchased scrap London Transport tube size vehicles and converted the Island's remaining railway between Ryde Pier Head and Shanklin for outside third rail operation. The stock was converted from fourth to third rail operation at the SR workshops at Wimbledon and transferred to the Island during 1966 and early 1967 for the 'New Railway' to take up operation on 20 March 1967. Two types of unit were formed; 4-car sets known as 4VEC (Class 485) and 3-car sets known as 3TIS (Class 486). In this illustration 4VEC No 044 poses for the camera on the Isle of Wight system.* British Railways

031
Private
Auxiliary

Above: *This view shows DMBS (Driving Motor Brake Second) No S11 of unit No 035, power is collected as in London Transport days via shoes mounted on both ends of this car, front end connections are sparse with only one brake pipe and one main reservoir pipe being provided. The three boxes above buffer beam level are electrical connections for control, brake and auxiliary circuits.* Colin J. Marsden

Below: *All units are formed the same way round with the DMBS (Driving Motor Brake Second) end being towards Shanklin. The driving cars are usually referred to by letter, by operating staff. 'D' ends are always the Shanklin end while 'A' cars are always facing the Ryde direction. These letters are carried on a plate just below roof height in the cab fronts. Unit No 035 with 'D' end near the camera pulls into Ryde Esplanade station during 1979* Colin J. Marsden

Above: *4VEC units are formed of two DMBS (Driving Motor Brake Second) cars and two TS (Trailer Second) cars, one of the TS cars being a former driving car and still retains its driving controls, but these are no longer used. Car No S31 currently in unit No 043 has de-icing equipment mounted in the redundant driving cab of the TS vehicle. Unit No 043 stands at Woking on the mainland prior to being shipped to the Isle of Wight Railway.* J. Bird

Below: *The striking external appearance of the stock can clearly be seen from this view of No S92, a trailer second car. Passenger access is via two double leaf air operated sliding doors (which are now painted silver). The TS vehicles are the oldest passenger stock in use on the British Railways standard gauge system being introduced first on London Transport in 1924.* Colin J. Marsden

Class 487

Class type: Waterloo & City
Car numbers: S51-S86
Introduced: 1940
Built by: English Electric
Equipment manufacturer: English Electric
Area of use: Waterloo & City Railway
Weight of cars: DMBS 23ton/TS 19ton
Length of cars: 49ft $1\frac{3}{4}$in
Height: 9ft 7in
Width: 8ft $7\frac{3}{4}$in
Brake type: Auto air
Nominal supply voltage: 600V DC
Type of power collection: Third rail
Traction motor type: English Electric 500
Available hp: 380hp (per m/c)
Number of traction motors: 2 (per power car)
Gear ratio: 69:16
Max unit speed: 35mph
Bogie type: Special design
Coupling type: Ward
Coupling restriction: Within cars only
Formation and seating: DMBS 40 second; TS 52 second
Note: Cars are not permanently coupled and can run in 1-, 2- or 5-car forms

Below: *One of the most unique railway systems in the country operates from Waterloo to Bank in the City of London. Trains are formed of tube sized vehicles of two types DMBS (Driving Motor Brake Second) and TS (Trailer Second), both are classified as 487. Trains are normally formed of five coaches during peak periods and two at off peak times. A 5-car formation with No S61 nearest the camera stands at Bank during late 1970s.* Brian Morrison

Above: *DMBS (Driving Motor Brake Second) coaches are single power cars with a driving cab at each end. Cars are capable of running singly or in multiple, power is collected from outside third rail collector shoes mounted on both bogies. DMBS No S58 stands at Alton whilst carrying out CM&EE trials during September 1977.*
Colin J. Marsden

Below: *Coaches are painted in standard rail blue livery with silver ends. On DMBS cars red headlights and tail lights are carried at all times. LT type couplings are provided with main reservoir and brake pipes situated above. Two DMBS cars stand at Waterloo Dept in this 1980 photograph. Trains are referred to by letters (A to E) by operating staff, these letters are carried on the front end emergency door.*
Colin J. Marsden

Above & below: *The trailer cars are usually marshalled in threes between two power cars and are used during peak periods. At other times trailer cars will be found stabled in the sidings at the Waterloo end of the line. No equipment is carried on trailer cars except for a Westinghouse triple valve brake unit. End connections are also sparse with only air pipes, control and lighting jumper socket being fitted. In the upper illustration car No S72 stands at Stewarts Lane Shed under repair, while in the lower photograph car No S76 stands in the sidings at Waterloo.* Both: Colin J. Marsden

Above & below: *Routine maintenance is carried out in the lines own workshops at Waterloo, but any major repairs or classified works overhauls are carried out at one of the Southern main line depots. Selhurst usually carry out general repairs but Stewarts Lane, Wimbledon and Eastleigh have been known to assist. In the upper view DMBS No S56 receives attention to its non-powered bogie at East Wimbledon EMU Depot during 1980. It will be seen that this vehicle has its nose end air pipe removed, this is done when stock is removed from the line by means of the hydraulic lift, clearances are very tight with even the couplings having to be removed prior to lifting and then being replaced when stock is at surface level. The lower illustration shows an unpowered bogie as fitted under the dummy end of DMBS and TS vehicles.* Both Colin J. Marsden

Class 491

Class type: 4TC
Unit numbers: 401-434
Computer numbers: 491401-491434
Introduced: 1966, 1974
Built by: BR/BREL York
Equipment manufacturer: GEC Traction Ltd
Area of use: Waterloo-Bournemouth/ Weymouth line
Weight of unit: 132ton
Length of unit: 265ft $4\frac{1}{4}$in
Height: 12ft $9\frac{1}{2}$in
Width: 9ft 3in
Brake type: Auto air/electro-pneumatic
Nominal supply voltage: From another unit/loco
Type of power collection: From another unit/loco
Traction motor type: None
Available hp: None
Number of traction motors: None
Gear ratio: None
Max unit speed: 90mph
Bogie type: B5
Coupling type: Buck-eye
Coupling restriction: SR powered units and Class 33/73 locos
Formation and seating: DTSO 64 second; TFK 42 first; TBSK 32 second; DTSO 64 second

Below: *Twentyeight 4-car TC (Trailer Control) sets of Class 491 were constructed at York works for use with the REP units on the newly electrified Bournemouth line in 1966 and 1967. Units are of identical appearance as the 4REP (Class 430) units except that the TRB (Trailer Restaurant Buffet) cars are replaced by TF (Trailer First) and the TBF (Trailer Brake First) are replaced by a TBS (Trailer Brake Second) vehicle. In this view unit No 417 leads a 12-car formation on a Waterloo to Bournemouth train near Wallers Ash.* Colin J. Marsden

Above: *Other factors that assist in recognition of TC units is that no third rail collector shoes are fitted. Driving cars are not fitted with underslung equipment and bogies are all of the B5 type. A further batch of six units of identical construction were introduced during 1974 and are recognisable from earlier members by the omission of lifting lugs situated at bogie centre position on the body sides. Unit No 433 passes Clapham Junction with a Waterloo to Bournemouth fast service during 1980* Colin J. Marsden

Right: *TC units often work in multiple with Class 73/1 and 33/1 locomotives by means of the high level jumper cables. In this view of 4TC No 409 coupled to power plant No 73.106, jumper connections can be seen; when coupled for this kind of operation power for heating and control is gained from the locomotive. The filler point under the cab side window on the TC unit is for window washing fluid.*
Brian Morrison

Class 501

Car numbers: See below
Introduced: 1957-1958
Built by: BR Eastleigh
Equipment manufacturer: English Electric
Area of use: Euston-Watford and Richmond-Broad Street
Weight of unit: 106ton
Length of unit: 181ft 9in
Height: 12ft 8in
Width: 9ft 6in
Brake type: Auto air/electro-pneumatic
Nominal supply voltage: 630V DC
Type of power collection: Third rail
Traction motor type: English Electric
Available hp: 740
Number of traction motors: 4
Gear ratio: 74:16
Max unit speed: 60mph
Bogie type: Mk 11
Coupling type: Screw
Coupling restriction: Within type
Formation and seating: M61135-M61188 DMBS 74 second; M70135-M70189 TS 92/94 second; M75135-M75189 DTBS 74 second

Below: *A fleet of 57 3-car units was constructed at Eastleigh during 1956/7 for use on the LMR London Division DC lines. Stock conforms to the basic 1957 SR requirements but are not fitted with buck-eye couplings. The livery currently applied is standard rail blue with full yellow ends. Units are identified by coach numbers rather than a set number and several sets are running misformed with odd and wrong cars. DMBS (Driving Motor Brake Second) No M61151 stands at Willesden Junction on a Watford working. Note: At the end of 1981 at least one unit had emerged from works painted in blue-grey livery.* Colin J. Marsden

Above: *The external appearance of the Class 501 cars is alway recognisable, as window guards are fitted across all drop light windows. In this view of DTBS (Driving Trailer Brake Second) the empty underframe can be seen, with only brake equipment carried. All cars are allocated to Croxley Green Depot and work on the Richmond to Broad Street and Euston to Watford lines.* Colin J. Marsden

Below: *The front end layout of the Class 501 units are immediately recognised as there is no other stock in service of the same design. Front end connections consist of (left to right in above plate) Control jumper sockets (2), Power jumper cable, Power jumper cable socket. Control jumper cables (2). A red tail indicator light is provided under the power jumper socket. Buffer beam connections consist of screw couplings and air pipes. DMBS (Driving Motor Brake Second) No M61142 approaches Broad Street during November 1980.* Colin J. Marsden

Class 503

Car numbers: See below
Introduced: 1938 and 1956
Built by: Metropolitan Cammell, BRCW Ltd
Equipment manufacturer: Metropolitan Vickers
Area of use: MerseyRail, Wirral Line
Weight of unit: 77ton
Length of unit: 176ft 11in
Height: 11ft 10$\frac{1}{4}$in
Width: 9ft 11in
Brake type: Auto air/electro-pneumatic
Nominal supply voltage: 650V DC
Type of power collection: Third rail
Traction motor type: BTH
Available hp: 540
Number of traction motors: 4
Gear ratio: 63:16
Max unit speed: 70mph
Bogie type: Single bolster
Coupling type: Screw
Coupling restriction: Within type only
Formation and seating: M28371-M28690 DMBS 56 second; M29702-M29846 TS 54 second; M29140-M29289 DTS 66 second

Left: *Although all classified as Class 503 stock this fleet of 43 3-car units can be divided into two parts. The first 19 being built by Metropolitan Cammell and Birmingham Railway Carriage & Wagon Company in 1938. Power collection for traction is via a slipper shoe mounted on both bogies of DMBS (Driving Motor Brake Second) vehicles and the outer end of DTS (Driver Trailer Second) cars. In this view No M29235M leads a 6-car train out of Bidston during early 1980.* L. Bertram

Above: *Front end communicating doors were added from 1972 as a Department of Transport directive as the stock would soon be required to work in tunnel sections of line. Air pipes are now duplicated either side of the front door, with control jumpers being housed under buck-eye level. In this view of DMBS (Driving Motor Brake Second) No M28675M underframe mounted power and control equipment can be seen together with the pick-up slipper mounted on the bogie.* P. Henk

Below: *Units are fitted with air operated sliding passenger doors of a double leaf type. 1938 built units have guard only operated doors while later 1956 built units have passenger controlled doors. This view of TS (Trailer Second) No M29710M has a modified drop light window system; only brake equipment is mounted under TS vehicles.* B. Edwards

Above: *Units of the 1956 build are recognisable from their older brothers by several minor details. The sole bar on 1956 stock is of the channel girder style, while on 1938 stock the sole bar is of the straight sided style. DTS No M29153M of 1956 build approaches Birkenhead Park during 1980.* P. Henk

Below: *Passenger push buttons for door operation are provided adjacent to double leaf doors on 1958 stock. No headcode system is carried by units but four white frontal lights and a destination blind are provided. At the rear, one of the bottom marker lights is covered by a red lens to act as a tail light. No M29131M arrives at Wallasey Village during 1980.* P. Henk

Above: *Units of the later build are equipped with a motor generator set mounted under the DMBS car, this provides auxiliaries at 50 volts, unlike the earlier units where auxiliaries were taken from the line volt supply (650v DC) and wired in series. DMBS No M28390M departs from Wallasey Grove bound for New Brighton on 8 August 1980.* L. Bertram

Below: *Another method of identifying the two different builds is by an additional bead stripe on 1938 stock at waist level, which was not built into later units. Most units are now endorsed with the Merseyrail PTE sign as well as the BR double arrow logo, DTS No M29131M stands in this illustration at Birkenhead North during 1980. Note this coach is minus shoe gear.* P. Henk

Class 504

Car numbers: See below
Introduced: 1959
Built by: BR Wolverton
Equipment manufacturer: English Electric
Area of use: Manchester-Bury
Weight of unit: 82ton
Length of unit: 133ft 3½in
Height: 12ft 7in
Width: 9ft 3in
Brake type: Auto air/electro-pneumatic
Nominal supply voltage: 1,200V DC
Type of power collection: Third rail
Traction motor type: EE
Available hp: 564
Number of traction motors: 2
Gear ratio: 63:17
Max unit speed: 70mph
Bogie type: Gresley design
Coupling type: Buck-eye
Coupling restriction: Within type
Formation and seating: M65436-M65461 DMBS 84 second; M77157-M77182 DTS 94 second

Below: *Class 504 2-car units were built during 1959 and 1960 for use on the Manchester Victoria to Bury line energised at 1200v DC. Power is collected from an outside third rail by means of an enclosed electrical shoe mounted on the bogie of each car. Unit design is of the classic 1959 Wolverton type and is almost identical in appearance to Classes 304, 305, and 308 units on the Eastern Region. In this view car No M77173 stands at Bury during 1979.* B. Denton

Above: *The front ends can be recognised from that of Class 304, 305 and 308 units in the following ways. Additional marker light and power jumper fitted under the driver's window, a small destination indicator is fitted on the roof in place of the headcode box; a centre cab window is fitted with car number painted above. These units do not carry set numbers and are referred to by coach numbers. DTS (Driving Trailer Second) No M77165 stands at Manchester Victoria during early 1981.* Colin J. Marsden

Below: *Recognition of the DMBS (Driving Motor Brake Second) coach and the DTS (Driver Trailer Second) coach is comparatively easy as on DMBS cars all traction and control equipment is mounted between the bogies, and the guard's van is positioned behind the driver's cab. If this plate and the one above are compared these differences will become apparent. DMBS No M65444 stands at Manchester Victoria during early 1981.* Colin J. Marsden

Class 506

Car numbers: See below
Introduced: 1954
Built by: Metropolitan Cammell, BRCW Ltd
Equipment manufacturer: English Electric
Area of use: Manchester-Glossop
Weight of unit: 105ton
Length of unit: 177ft 10in
Height: 12ft 8¼in (max)
Width: 9ft 3in
Brake type: Auto air/electro-pneumatic
Nominal supply voltage: 1,500V DC
Type of power collection: Overhead
Traction motor type: English Electric
Available hp: 432
Number of traction motors: 4
Gear ratio: 69:18
Max unit speed: 75mph
Bogie type: Single bolster
Coupling type: Screw
Coupling restriction: Within type only
Formation and seating: M59401-M59408 DMBS 52 second; M59501-M59508 TS 62 second; M59601-M59608 DTS 60 second

Above right: *In this view the DTS (Driving Trailer Second) end leads as a Class 506 set approaches Manchester Piccadilly. Underframe mounted equipment on trailer vehicles is confined to brake control systems. Front end connections consist of one control jumper mounted at waist height. Air connections are fixed at buffer beam level.* Colin J. Marsden

Right: *During the summer of 1980, one unit (Coach Nos M59404M, M59504M, M59604M) emerged from the works painted in blue and grey livery which has given a more pleasing look to these rather mundane units. Underframe mounted equipment is confined to the DMBS (Driving Motor Brake Second) vehicle. This newly painted unit stands at Hadfield during September 1980. Note: By the end of 1981 the majority of units were repainted in inter-city colours.* Brian Morrison

Below: *This fleet of eight electric units provide the only EMU power using the 1,500v DC overhead system. Units were constructed in 1954 and are currently allocated to Reddish Depot. DMBS (Driving Motor Brake Second) cars are easily recognisable from DTS (Driver Training Second) cars by having the roof mounted cross armed pantograph fixed on the roof above the cab. Units are not allocated set numbers, but all cars with numbers ending in the same two digits usually run together. No M59402M stands at Manchester Piccadilly in this early 1981 photograph.* Colin J. Marsden

M59402M

M59404M

Class 507

Unit numbers: 507001-507030
Introduced: 1978
Built by: BREL York
Equipment manufacturer: GEC Traction Ltd
Area of use: Liverpool-Garston and Southport
Weight of unit: 98ton
Length of unit: 199ft 7in
Height: 11ft 9in
Width: 9ft 3in
Brake type: Electro-pneumatic/dynamic
Nominal supply voltage: 660-750V DC
Type of power collection: Third rail
Traction motor type: GEC G310AZ
Available hp: 880
Number of traction motors: 8
Gear ratio: 69:14
Max unit speed: 75mph
Bogie type: BX1
Coupling type: Tightlock
Coupling restriction: Within unit type
Formation and seating: BDMS 74 second; TS 86 second; DMS 74 second

Below: *New stock for Liverpool's Northern Line was delivered from late in 1978, when 33 3-car units of 1972 high density design were constructed at BRE York. In external appearance units are the same as Class 313, 314, 315 and 508 units in use on the Eastern, Scottish and Southern Regions. Units are fitted with sliding doors which are under the control of the guard. Power is collected via bogie mounted collector shoes fixed to the bogies of driving vehicles. Unit No 507030 with BDMS (Battery Driving Motor Second) leading stands at Southport during late 1980.* B. Edwards

Above: *All units are currently allocated to Hall Road Shed in Liverpool and are used on the Southport to Liverpool line. Front end connections are of the same layout as that of 508 stock (for explanation of front end connections see Class 508 section). Over the next few years the Southern Region's 508 units will be allocated to the Liverpool area and work alongside the 507 units. No 507026 in this illustration stands at Waterloo station on the Southport to Liverpool line.* B. Edwards

Below: *All underslung traction and control equipment is housed between the bogies of the BDMS and DMS coaches. Along the upper part of the coach body above the two centre windows is situated a yellow door indicator light, which when illuminated denotes that the doors on that vehicle are open or in release mode, this light is also illuminated if the passenger communication apparatus is activated. Unit No 507013 nears Sandhills on a Southport to Liverpool working.* B. Denton

Class 508

Unit numbers: 508001-508043
Introduced: From 1979
Built by: BREL York
Equipment manufacturer: GEC Traction Ltd
Area of use: SR/SWD Suburban lines
Weight of unit: 122ton
Length of unit: 264ft 10in
Height: 11ft 9in
Width: 9ft 3in
Brake type: Electro-pneumatic/dynamic
Nominal supply voltage: 660-750V DC
Type of power collection: Third rail
Traction motor type: GEC G31/AZ
Available hp: 880hp
Number of traction motors: 8
Gear ratio: 69:14
Max unit speed: 75mph
Bogie type: BX1
Coupling type: Tightlock
Coupling restriction: Within class only
Formation and seating: DMS 74 second; TS 86 second; TS 86 second; BDMS 74 second

Left: *From late in 1979 Class 508 units began to enter service on the suburban lines out of Waterloo on the Southern South Western Division. A total of 43 4-car units were built at BREL York of the standard 1972 high density style. No 508006 in this illustration approaches Wimbledon on a Waterloo-Shepperton working during 1980.* Colin J. Marsden

Above & below: *This fleet of units are of the same general body appearance to that of the Classes 313, 314, 315, and 507 types, but have minor identifying factors. The two driving cars are of identical design except that one end (the B end) carries batteries, but this cannot be seen from the outward appearance. Driving cars are classified DMS (Driving Motor Second) and BDMS (Battery Driving Motor Second). Power is collected via shoes mounted on the outer ends of DMS and BDMS cars. In the upper illustration unit No 508031 shows the DMS end, equipment carried between the bogies consists of coach heating and control equipment. In the lower illustration the BDMS end of No 508031 is shown.* Both Colin J. Marsden

4
via Kingston
Strawberry Hill
508 006
1
2
Class 508
3
2
1
5
8
6
7

Left: *Class 508 front end layout: (**1**) White marker lights, (**2**) Red tail indicator lights, (**3**) Unit headlight, (**4**) Air warning horns, (**5**) Buck-eye coupling, (**6**) Electric/air connection box, (**7**) Drum switch unit, (**8**) Manual uncoupling system. Unit No 508006 was photographed at Strawberry Hill during early 1980. This configuration of equipment also applies to Class 313, 314, 315 and 507 stock.* Colin J. Marsden

Above: *The tight lock type of coupling is fitted to all breeds of 1972 type high density stock, the upper part of the coupling unit is the basic buck-eye while the lower part contains all electrical and air connections. During coupling the cover of the electrical and air connections automatically drops away as the units close together. 1972 design high density units are not compatible to any other type of unit for multiple operation, but stock can be assisted by buck-eye fitted units or locomotives in the case of emergency.* Colin J. Marsden

Below: *508 units are recognisable from Class 507s in that they are fitted with passenger operated doors activated by push buttons mounted either side of the double leaf doors, unlike 507 stock. The front end grab rails either side of the emergency door have been removed on 508 stock due to fouling of assisting stock in the case of emergency. Unit No 508014 stands alongside 4VEP (Class 423) unit No 7742 at Wimbledon during the summer of 1980.* Colin J. Marsden

Departmental Stock

1972 Design High Density Stock

Left: *The original 1972 high density stock built at York and allocated to the Southern were classified as 4PEP (Class 461) units. After finishing their evaluation on the Southern the two 4-car and one 2-car units were allocated to the Research Centre at Derby. Unit No 4002 in the upper plate stands outside the Research Centre during 1979, painted in an experimental livery.* Colin J. Marsden

Below left: *From time to time unit No 4002 has returned to the Southern for various rolling stock tests, usually involving bogies. When in transit from Derby to the Southern Region the unit is split and coupled cab to cab as no other rolling stock or locomotive is compatible to its coupling. No 4002 stands coupled cab to cab at Clapham Junction in this illustration. Note different bogies under each coach.* Colin J. Marsden

Above: *One of the unique features of 'PEP' units was that each axle of every coach was powered, and this gave a rapid acceleration rate. Driving cars were provided with two pairs of double leaf passenger activated air assisted sliding doors on each side, while on non driving cars three sets of doors were provided each side. MSO (Motor Second Open) No S62426 is nearest the camera in this view of unit No 4002 at Clapham Junction.* Colin J. Marsden

Below: *The 2-car unit No 2001 after withdrawal was also utilised at the BRB Research Establishment at Derby. Strengthened to three cars and renumbered 920001 it was put into technical service on the Eastern Region being later used in Scotland. At the time of writing No 920001 is at the Derby Technical Centre. In this view MBSO (Motor Brake Second Open) No ADB975432 formerly No S64301 stands at Derby along with other cars of PEP stock.* Colin J. Marsden

Above: *PEP units were fitted with a new and experimental* Scharfenberg *coupling, which incorporated not only a physical coupling but air and electric connections as well. In this view the physical coupling can be seen in the centre while at the edges two vertical drum rollers hide the electrical connections.* Colin J. Marsden

Below: *The bogies fitted were also of an experimental type and were classified as Mk 6. In this view the bogie can be seen fitted with its third rail collector shoe mounted on a radial shoe arm.* Colin J. Marsden

Class 97
EMU Locomotives

In connection with various electrification projects during the 1970s several Class 501 DMBS (Driving Motor Brake Second) cars were converted into battery electric locomotives. During conversion work all internal seating was removed, roofs were lowered and flattened, front ends were completely rebuilt, and all doors were sealed up, with louvres fitted into the window bays. In the plate on the left the front end can be seen painted in yellow/black wasp colours, MU jumper equipment can also be seen fixed under the driver's window. The plate below clearly shows the flattened roof of the vehicle and the sealed up doors.
Both Colin J. Marsden

Above: *Power for these units is provided by traction batteries carried in the former seating bays, conversion work was carried out at BREL Doncaster Works. These battery locomotives are numbered in the departmental locomotive number series viz: 977XX. In this illustration No 97703 formerly No M61182 stands at Doncaster after conversion.* Colin J. Marsden

Below: *These battery units usually run in 2-car formations and are currently allocated to Birkenhead North, Hornsey, Glasgow Shields Road and Cricklewood Sheds, and are used for hauling and working with electrification engineers' trains. In this illustration Nos LDB 975407 and LDB 975408 soon to be renumbered 97707 and 97708 stand at Hornsey during 1975.* British Railways

Southern Region Stock

Above & below: *To cater for depot stores requirements, the Southern Region operates a fleet of three 2-car stores units, which operate a weekly cycle covering all depots and Eastleigh BREL works. Units Nos 022 and 023 are formed of redundant MBS (Motor Brake Second) coaches of former SR 2HAL units, and No 024 is formed of MBS coaches of redundant SUB units. The coaches are stripped of all seating, and fitted with open pallets. During rebuilding work rollerblind headcodes and high level air pipes were fitted, this enabling units to work with buck-eye fitted stock, albeit with through brake connections only. In the upper illustration unit No 022 stands at East Wimbledon depot, in the lower illustration unit No 024 stands at Hither Green shed during 1978. Note the shoe protection boards.* Both Colin J. Marsden

Above: *Another of the former Class 501 cars has seen further use after withdrawal from passenger service, DTBS (Driving Trailer Brake Second) No M75165 was converted for use as a Southern Region research coach and allocated running number 975032. The coach is currently undergoing tests based at Strawberry Hill Depot. Front end connections consist of low level air pipes, train heating connections, waist level air pipes, 27 way control jumper and power jumper. A further jumper socket is housed below the headcode box which is used for test purposes. No 975032 in this illustration stands coupled to a DMBS vehicle of a 4CEP Class 411 unit at Strawberry Hill during early 1980.* Colin J. Marsden

Below: *During 1959/60 four former 4SUB trailers were converted to act as de-icing trailer cars, being able to lay fluid and scrape the live rail, but not propel themselves. Trailer de-icing cars are usually marshalled between two 2-car 1951 or 1957 series units and are fitted with high level jumpers at each end. Live rail observance ports can be seen protruding at each end of the coach on the body side. Cars are currently allocated to Eastleigh and Ramsgate Sheds.* J. Scrace

Above & below: Three SR de-icing units were taken into stock during 1967/8 and numbered 001-3, when redundant power cars of former 2HAL units were converted. Units were formed of two power cars, each fitted with spraying equipment. During the 1980s units were fitted with Sandite rail cleaning systems to improve track adhesion. Only two of the three original units still remain in service (001 and 003). In the upper illustration unit No 001 stands at Bournemouth depot during 1977, the lower plate shows the bogie and shoe equipment.
Both Colin J. Marsden

Above & below: *During the late 1970s the need for a newer de-icing unit could be seen and a major rebuilding programme was undertaken. The units chosen for conversion were former 4SUB units. Conversion work included the sealing up of the driver's doors and giving access to the cab via the adjoining brake van, sealing up all but two doors in the coaches former seating bays, removing all seats and internal equipment and refitting with new fluid tanks and control systems. The front end of these units is hardly recognisable as being that of a SUB unit, new slimmer windows, roller blind route indicators and high level air pipes were fitted. In the illustration above unit No 008 stands at Wimbledon Park Depot during late 1980, showing the body side appearance. The lower illustration shows the new front end layout of unit No 009 standing alongside Class 508 number 508032.* Both Colin J. Marsden

Above: *Southern Region departmental unit No 055 was introduced during 1974 when a former 4SUB No 4367 was converted into a mobile instruction unit. All seating and internal fittings were removed, and each coach was fitted out as a classroom; during conversion work gangways were provided between cars. No 055 is usually to be found at either Selhurst, Chart Leacon or Wimbledon Park sheds. In this view taken during late 1979 No 055 stands at Wimbledon.* Colin J. Marsden

Midland Region Stock

Below: *To assist with rail adhesion on the Liverpool third rail system, two DMBS coaches from withdrawn Class 502 units have been coupled back to back and fitted with rail cleaning equipment. The units now carry departmental running numbers ADB977017/8 and are allocated to Hall Road Liverpool. In this illustration the two car set stands at Southport coupled to Class 25 No 25.192 during Septmber 1981.*
Brian Morrison

Above: *The Southern Region General Manager's Saloon has been included in this volume as it is fully compatible to SR EMU stock as well as Class 73/1 and 33/1 locomotives. The coach No TDB975025 was converted during 1969 from a former Hastings line buffet car No S60755. The coach is normally kept at Stewarts Lane Depot when not in use. The car is fitted with low level air pipes, train heating cables, waist level air pipes and 27 way control jumpers. Just below roof height three powerful spot lights have been fitted being mainly used for tunnel inspections. A standard two digit SR headcode is fitted under the centre window. Internally the coach is built with an observatory at each end with guard and cooking facilities amidships. An interesting feature of this coach is that standard 1963 style driving controls are fitted at each end. No TDB975025 stands parked at Stewarts Lane during late 1978.* Colin J. Marsden

Class 317

At the time of going to press a fleet of new generation emu vehicles is under construction at BRE Derby and York works, which when introduced during 1982 will enter service between Bedford-St Pancras and Moorgate. The only data available on these units is given below.

Numbers: 317.301-317. 348
Equipment manufacturer: GEC Traction Ltd
Nominal supply voltage: 25Kv ac/ 750v dc
Control supply voltage 110v
Type of power collection: Overhead/3rd rail
Formation and seating: DTS 74 second; MS 79 second; TC 22 first, 46 second; DTS 74 second